Magic and Warfare

Magic and Warfare

Appearance and Reality in Contemporary African Conflict and Beyond

Nathalie Wlodarczyk

palgrave
macmillan

MAGIC AND WARFARE

First published in 2009 by
PALGRAVE MACMILLAN®
in the United States—a division of St. Martin's Press LLC,
175 Fifth Avenue, New York, NY 10010.

Where this book is distributed in the UK, Europe and the rest of the world, this is by Palgrave Macmillan, a division of Macmillan Publishers Limited, registered in England, company number 785998, of Houndmills, Basingstoke, Hampshire RG21 6XS.

Palgrave Macmillan is the global academic imprint of the above companies and has companies and representatives throughout the world.

ISBN: 978–0–230–62102–2

Library of Congress Cataloging-in-Publication Data

Wlodarczyk, Nathalie.
Magic and warfare : appearance and reality in contemporary African conflict and beyond / Nathalie Wlodarczyk.
p. cm.
Includes bibliographical references and index.
ISBN 978–0–230–62102–2
1. Sierra Leone—History—Civil War, 1991–2002—Religious aspects. 2. Military art and science—Sierra Leone. 3. Magic—Political aspects—Sierra Leone. 4. Shamans—Sierra Leone. 5. War and society—Sierra Leone. 6. Ethnic conflict—Africa—Case studies. 7. War—Religious aspects—Case studies. 8. Military art and science—Africa—Case studies. 9. Magic—Political aspects—Africa—Case studies. 10. War and society—Africa—Case studies. I. Title.

DT516.826.W55 2009
966.404—dc22 2009013783

A catalogue record of the book is available from the British Library.

Design by Newgen Imaging Systems (P) Ltd., Chennai, India.

First edition: November 2009

10 9 8 7 6 5 4 3 2 1

Printed in the United States of America.

Always use the word "Africa" or "Darkness" or "Safari" in your title. Subtitles may include the words "Zanzibar," "Masai," "Zulu," "Zambezi," "Congo," "Nile," "Big," "Sky," "Shadow," "Drum," "Sun," or "Bygone." Also useful are words such as "Guerrillas,'" "Timeless," "Primordial," and "Tribal." Note that "People" means Africans who are not black, while "The People" means black Africans.

Binyavanga Wainaina, "How to Write about Africa,"
A View from Africa (London: Granta, 2005)

One would have hoped that by now we would have broken the habit of identifying those traits which seem to make us ostensibly unlike others—the color of our skin, the language we speak, the food we eat, the beliefs we espouse—and come to terms with what all human beings have in common, for better or worse, and seeing beneath the surface of cultural differences, comparable imperatives, logics and dispositions.

Michael Jackson, *In Sierra Leone*
(Durham, NC: Duke University Press, 2005)

Contents

Part III

Map, Table and Figures

ACKNOWLEDGMENTS

This book would not have been possible without the support and help of a number of people.

In London, Christopher Coker first raised the issue of witchcraft and magic in 2000 and put me on the path toward this book. Dominique Jacquin-Berdal and her passion for Africa sparked my own infatuation with the continent and was an enormous inspiration. On a visit to Aberystwyth, Mike Williams introduced me to Bourdieu's work on practice and took the time to listen to my then rambling thoughts on function and functionalism. Together, they helped lay the foundation for this book.

Once on the path, Jan Willem Honig and Mats Berdal helped me focus my thoughts and pull a multifaceted project together into what I hope is a meaningful whole. I am very grateful to them for helping guide a project that did not fit neatly into any one discipline and appreciating the value of exploring something slightly off the beaten track.

In Sierra Leone, Mariama Konneh patiently guided me through my first encounter with the country and continued to be a pillar of support and a great friend on all my visits as well as in London. She also introduced me to key people that helped make the project alive and relevant. Together with Vamba Konneh she opened their home to me and always made me feel welcome. James Roscoe took me in when I first arrived in Freetown, repeatedly put me up in his home on my visits and was a most gracious host. Cobus Claassens introduced me to a CDF network that proved crucial to my research. Michael Josiah was invaluable in his support both in Freetown and on the road, and helped me with translations as well as to understand some of the subtleties of the CDF that would likely have passed me by without our long conversations. Michael was also extremely patient with me despite my limitations for which I am very grateful. Andy Ianuzzi and Bianca Suciu smoothed the way through the red tape at the Special Court for Sierra Leone and gave very welcome and thought provoking perspectives on the CDF. Rebecca Ehret invited me into the community of researchers in Freetown despite my sometimes short visits and gave me the confidence to present my work to anthropologists. Karen Barnes was a great travel companion and together with Rosalia Gitau helped me take my mind off work when needed.

To complement my research in Sierra Leone I am very grateful to Krijn Peters and Paul Richards who let me use their unpublished database of interviews with ex-combatants in Sierra Leone. Danny Hoffman kindly sent me his thesis and allowed me to use one of his very powerful photos on the cover of this book.

I have been lucky to be surrounded by friends and family that have commented on and criticized my work, helping to make it better. At King's College, Dawda Jobarteh, Ivan Zverzhanovski, Tanja Schumer, Alisa Carrigan, and Jessica Lincoln have been a great source of support and always willing to put the world to right over a drink. I am very grateful to all of them for their encouragement and optimism about my work. Martin Kimani encouraged my excitement about Bourdieu and patiently read what I came up with. OB Sisay and Thoko Kaime were inspirational and hugely supportive and optimistic, making me believe that I had something useful to say about both Africa and Sierra Leone. Kate Gibson and Eugenia Zorbas have been great friends throughout this process and have always been a source of emotional as well as professional support. Together with Patrick Cullen, Eugenia commented extensively and substantively on my early drafts and saw me through the conceptualizing of the project when I first started. They have both helped make the final product far better than it would have been without their constructive input. I also received some very valuable and welcome steers from Chris Ruane and Sigbrit Franke who both took on the first full draft of the manuscript. For the final stretch I am eternally grateful to Malte Gerhold for giving me the confidence to finish and seeing me through the emotional stress of the last months and weeks.

My parents, Lena and Michel Wlodarczyk have encouraged me throughout my life and this project was no exception. Even when the years accumulated they stood by me and offered all the love and support I needed to see it through. Without them this book would never have been possible. Although he did not know it, my grandfather André Casimir Wlodarczyk, made my first trip to Sierra Leone possible at a time when it was looking tricky. He is not here to see the final result, but I hope he would have liked it.

Finally, this book also owes its existence to the School of Social Science and Public Policy at King's College London, the Department of War Studies, and the University of London Central Research Fund, who all helped fund my research in Sierra Leone.

In one way or another everyone mentioned has helped me complete this book, and for their support and contributions I will always be grateful. The faults of course remain entirely my own.

ACRONYMS

AFRC	Armed Forces Revolutionary Council
APC	All People's Congress
CDF	Civil Defence Forces
ECOMOG	ECOWAS Monitoring Group
ECOWAS	Economic Community of West African States
EO	Executive Outcomes
HSM	Holy Spirit Movement
HSMF	Holy Spirit Mobile Force
INPFL	Independent National Patriotic Front of Liberia
LRA	Lord's Resistance Army
LURD	Liberians United for Reconciliation and Democracy
NPFL	National Patriotic Front of Liberia
NPRC	National Provisional Ruling Council
NRA	National Resistance Army
OBHS	Organised Body of Hunting Societies
RSMLF	Republic of Sierra Leone Military Force
SLA	Sierra Leone Army
SLPP	Sierra Leone Peoples Party
ULIMO	United Liberation Movement for Democracy in Liberia
UNLA	United National Liberation Army
UN	United Nations
ZANLA	Zimbabwe African National Liberation Army

Brief Chronology of the Sierra Leone Civil War (1991–2002)

1991	The Revolutionary United Front (RUF) launch their insurgency from across the Liberian border
1992	The National Provisional Ruling Council (NPRC) depose the government of Joseph Momoh in a coup
1995	The NPRC hire private military contractors Executive Outcomes to help them fight the rebels
1996	Elections are held and Ahmed Tejan Kabbah is elected president together with a Sierra Leone People's Party (SLPP) government
	The government and the RUF sign the Abidjan peace accord extending an amnesty to the rebels
1997	Junior officers oust Kabbah in a coup and establish the Armed Forces Revolutionary Council (AFRC) under the leadership of Johnny Paul Koroma. President Kabbah flees to Guinea.
	Nigerian peacekeepers in Sierra Leone are re-mandated as ECOMOG II
1998	Kabbah is reinstated as president through the joint efforts of ECOMOG and the Civil Defence Forces
	The UN establishes an observation mission for Sierra Leone, UNOMSIL
1999	AFRC/RUF forces enter Freetown and causes widespread destruction and killing 5,000 people
	The Lomé Peace Agreement is signed by the government and rebels
	The UN establishes a new mission for Sierra Leone, UNAMSIL. 6000 troops are authorized for peacekeeping

2000 UNAMSIL troops are increased to 11,000 in February and in May to 17,000

ECOMOG withdraws from Sierra Leone

British Special Forces intervene to free UN peacekeepers taken hostage by the RUF

2001 Disarmament begins

2002 Kabbah wins a second term in elections and declares the war over

Brief Chronology of the Civil Defense during Its Main Mobilization Period (1992–1998)

1992 *NPRC coup*
civil defense units are supported by the NPRC to boost local defense

1993 By the end of 1993 at least three traditional hunting guilds are actively assisting the SLA

1994 Civil Defense relations with the SLA sour as the rebel-soldier collaboration spreads

1995 Kamajor initiation ceremonies begin in Kale, Bonthe district

1996 *Elections*
Kamajors overrun the RUF headquarters Zogoda, Kenema district
Large-scale initiations into the Kamajor society begins across the southeast
Chief Hinga Norman, leader of the Kamajor civil defense militia, is appointed National Coordinator for Civil Defence and Deputy Minister of Defense by newly elected President Kabbah

1997 The Civil Defence Forces (CDF) are given official status
AFRC coup
The CDF reject a call from the junta to demobilize and disarm, and instead regroup in their strongholds
Talia Yawbeko becomes the main CDF base known as Base Zero
Operation Black December is launched targeting access roads in the south-east to cut rebel supplies and reinforcements

1998 The AFRC/RUF are driven out of Freetown by ECOMOG and CDF in late February

The CDF are placed under the control of ECOMOG by President Kabbah in March

Kabbah formalizes the CDF's position as part of the Armed Forces in September

Map 1 Sierra Leone

CHAPTER 1

INTRODUCTION

> You call them medicine men, but you have your own armor in western armies and sometimes it does not work, which is the same with us. Sometimes we get hurt and sometimes we get killed, but that doesn't mean it doesn't work.[1]

In 2003 the Special Court for Sierra Leone indicted those it held to be most responsible for war crimes during Sierra Leone's civil war. The two rebel leaders of the insurgency and their patron were, however, beyond reach for the court. Foday Sankoh, leader of the rebel Revolutionary United Front (RUF), had died in prison before the indictments were made. Johnny Paul Koroma, head of the military junta that had collaborated with the RUF, the Armed Forced Revolutionary Council, was missing and presumed dead after escaping custody in Freetown. The President of Liberia and backer of the RUF, Charles Taylor, was only indicted later and was protected for several years first as head of state in Liberia and later in protective exile in Nigeria.[2] With the rebel culprits dead or untouchable, the indictments of their enemies—the pro-government Civil Defense Forces (CDF)—who were all at hand, received widespread attention. They were particularly noted as one of the indictees, Sam Hinga Norman, was a government minister and a man considered by many to be a national hero.[3] But one of the other indictees also raised some eyebrows amongst Western observers for quite different reasons—Allieu Kondewa was the High Priest of the CDF and its main group the Kamajors. He had been responsible for the ritual initiation and magical protection that the force had become associated with on the ground. It appeared, then, that the court mandated to provide justice to a country ravaged by the most brutal civil war, and for whom the three leaders of the insurgent forces were beyond reach, was instead going to try a national hero and an illiterate witchdoctor.

On some levels the idea of a witchdoctor being involved in the war and its often bizarre manifestations was perhaps not so strange to an outside world that had followed the conflict cursorily through news reports of mutilations, cannibalism, and a reliance by fighters on magical charms. But the indictment certainly cemented this view of the war as a peculiarly African type of war where superstition and AK-47s combined to wreak havoc on a country and people.

MAGIC, CULTURE, AND THE NEW WORLD ORDERS

The civil war in Sierra Leone began in 1991, at the dawn of a new era. In the early 1990s, academic attention in the generally cold war-centric discipline of International Relations shifted toward the developing world and the challenges brought by the end of the cold war. With the grand ideological struggle of the twentieth century perceived to be resolved in the West's favor and the coming to an "end of history," other problems in the international realm could be given greater consideration.[4] With worries of a third world war alleviated by the fall of the Iron Curtain, the civil wars that tore apart many African states in the 1990s rightly received significant attention. After having been largely ignored by mainstream political commentators and theorists for the duration of the cold war, Africa became a test case for theories attempting to explain the violence that quickly came to define the post-cold war era. As it happened, the end of superpower struggle did not lead to peace and stability, rather the opposite. Although brutal civil wars erupted also in Europe and parts of the disintegrating Soviet Union, in Africa they were both more numerous and shocking in their brutality. By the middle of the 1990s civil wars were raging in Liberia, Sierra Leone, Somalia, Mali, Niger, Sudan, Algeria, Angola, Rwanda, Burundi, and Uganda. What was then Zaire would soon follow, as would Ethiopia and Eritrea.

In the 1990s Africa therefore came to represent war, but also apparent chaos as it seemed as though the removal of superpower support for ideologically driven proxy wars had instead unleashed a return to primitive tribal wars. The presence of spiritually imbued tactics on the battlefields across the continent did not do much to dispel these interpretations. In Liberia and Sierra Leone fighters claimed to be immune to bullets thanks to protection awarded through rituals—the kind of protection that Allieu Kondewa later stood trial for administering. In Uganda, the Lord's Resistance Army claimed to be guided by the Holy Spirit that possessed its leader Joseph Kony and in almost all the other wars on the continent rhetoric referencing spirits or amulets and potions turning bullets into water could be found. Clearly, there were elements to these wars that could not be easily explained by or fit into the traditional conceptions of warfare taught in war colleges and academic institutions around the world. So what did it mean?

Observers rose to the challenge with varying degrees of perceptiveness. Some opted for explanations anchored in the idea that African states were regressing to a violent tribal past after the civilizing influence of both colonialism and super power domination had been removed. These fit with attempts to redefine world order that relied on the idea that the world was rapidly crumbling from environmental degradation and population explosions, creating an economically divided and deeply unstable world of small elites and deprived masses. In this world, fragmented states and populations would come to blows over scarce resources and resort to past methods of warfare as the modern world became increasingly inaccessible.[5] Others attempted more nuanced interpretations that looked for the rational motivation behind the violence to explain why in many cases it simply made sense, relying on less deterministic explanations of economic marginalization, changing interstate power-relations, reemerging political grievances and entrepreneurial criminal activity.[6] Whereas these explanations could make sense of brutality as expedient and cheap they failed to explain the use of magic beyond a crude tactic for intimidation. The early post-cold war years also heard voices pleading for a better understanding of culture as a determinant of behavior in warfare. Building on the thinking of pioneers in the championing of culture in strategic studies like Adda Bozeman, a number of academics again pointed to culture as a neglected area of study and perhaps one that could help explain what was happening in wars in Africa and elsewhere.[7] But culture is a slippery slope that can just as easily become a tool for superficial analysis, where all the more difficult to explain behavior gets categorized as "culture," as it can offer real insights into the subtleties of particular acts or ideas. In the early hours of the new millennium the former gained a steady grip on much mainstream analysis and policy.

The attacks on New York and Washington DC in September 2001 rapidly swung attention away from the consideration of local agendas and interrelated and overlapping motivations toward a sharper dichotomy dividing the world into territories hosting Islamic terrorists and territories prone to being infiltrated by Islamic terrorists. In turn, this view came to equate certain types of conservative Islam with a violent agenda, effectively putting them into the "culture" box and linking it to an emerging threat—if these strands of conservative Islam were present in a country it was assumed to be a breeding ground for terrorists and a source of international destabilization. Africa, where the conservative Salafi and Wahhabi strands of Sunni Islam have been very influential, suddenly appeared as a frontline in the fight against jihadi terrorism.[8] Despite the fact that Christianity and Islam have coexisted largely peacefully on the continent and in most places continue to do so, Africa became interesting in relation to the role it might play in the global "war on terror."[9]

While this has no doubt been a distraction from some of the analytical undertakings in the early post-cold war years, it also emphasizes the ever pressing need of differentiating between motivations that drive behavior—whether to correctly identify the relative threat of terrorist activity or to assess the most suitable means of engaging with armed groups to promote peace. Whereas in the 1990s the use of witchdoctors on African battlefields had bemused external observers, the appearance of a spiritual agenda in violence perpetrated on a large scale in the West was perceived very differently. As a result, studies of various strands of Islam to understand some of the violent manifestations that claim to have grown out of them have increased exponentially. Intelligence organizations in the West that used to have basements full of Russian speakers, are now trading them in for those fluent in Arabic and Pashto. There is, however, a risk that yet a metanarrative will yet again capture the analytical space that considers conflict by focusing too closely on one influence (in this case the more violent interpretations of Salafist Sunni Islam). Looking at the dynamic through which other belief systems influence strategies and tactics therefore provides both new specific insights and a test case for theories developed specifically to address the "War on Terror" and its analytical narrative.

It would be disingenuous to say that all study of African conflict has been captured by the War on Terror debate. Study of the conflict and development nexus has continued to flourish since its first dramatic push in the 1990s. This is a nuanced field that explicitly seeks to connect social, economic, and political dynamics to understand how they impact on security. But much as with the work on economic agendas in conflict that sought to disprove theories of chaos and anarchy as the underlying drivers of conflict in the developing world, this strand of research tends to focus almost exclusively on the spheres of economic calculations and power structures. To complete this picture, which is seeking to provide a more holistic perspective on conflict, we need to add an improved understanding of the ideas and beliefs that inform calculations about gain and power. If we do, we will be better equipped not only to map conflicts in sub-Saharan Africa but the drivers behind individuals and groups that turn to terrorism, whether at home or abroad.

Bringing "Magic" into Strategic Studies

Religious belief systems of any kind sit uncomfortably in political science and even more so in strategic studies. Religious aspects of conflict (or influence on politics) generally receive more skeptical and cynical treatment than other dimensions, be they economic, social, or ideological. Firmly anchored on the path of thinking grounded in the Enlightenment, Western elites—academics and in public life—tend to relegate spirituality to a prior developmental stage and a sign of primitive intellectual development. Even in the corridors of power where religious belief is vocally referenced in relation to policy decisions, as was the case in the White House in the run-up to the Iraq war, the designers and implementers (and certainly analysts) of policy tend to lean firmly in the other direction. This is even more the case with the practical application of spiritual convictions. Strategic theory as a field of study is the consideration of the recurring patterns of thought and behavior that hold war together as a recognized activity beyond the basic use of lethal force by opposing sides. In many cases it also deals with the relative effectiveness of different variations of these patterns. Strategic theory incorporates assumptions about the rationality of behavior and is largely concerned with the means-end relationships that make up war. The logic involved in devising appropriate means for chosen ends is seen as instrumentally rational—the means will be chosen to maximize the chance of achieving the ends. When considering outcomes, it is therefore not surprising that contemporary strategic theorists generally steer clear of factors that are not easily quantifiable and stick to measurable material influences when commenting on strategy and tactics. However, the internal logic in the process of a means-end relationship can be influenced and directed also by factors that are less amenable to quantification, such as religious conviction. The challenge is to connect in a meaningful way the instrumental rationality of a means-end process with supposedly irrational influences. In other words, identifying the instrumental rationality—or function—of allegedly irrational activities. Without incorporating the influence of these activities into our theories about war they will inevitably be less useful as tools that explain or even direct warfare.

The benefits of making way for a broader set of influences on warfare in the theories we use to assess it are many. For one, it will help us understand why an armed group like the Kamajors in Sierra Leone kept the use of magic at the very core of their movement. It will also help guide analysis in other places where the use of spiritually imbued rituals find their way onto the battlefield. But beyond this it will help us analyze behavior in warfare more broadly. In today's world the spread of Western values around the globe has arguably been significantly slowed by increased competition both from new emerging global players and resurgent local ones. Whether it is the Bolivarians in Latin America, the Salafists in the Middle East and Asia or the opportunities to opt out of Western models provided by dramatically increased Chinese investment and involvement in Africa, the global normative space is changing. Engaging with this world will require a willingness to understand local, or simply different, perspectives. This is challenging enough in peacetime, but when hostilities break out both the challenge and the imperative to face it increase significantly. Whether tackling al-Qaeda in the Islamic Maghreb in Algeria, al-Shabab in Somalia, the Tamil Tigers in Sri Lanka or the Justice and Equality Movement in Sudan, a solid understanding of the group's internal dynamic is required to make that engagement successful. This dynamic will be shaped by the shared beliefs and aims of the group, the challenge they feel they are facing up to and the resources available to them to pursue that agenda.

This brings us back to our ritual expert and by now convicted war criminal, Allieu Kondewa.[10] A ritual expert who becomes so central to an armed group that he ends up being indicted for war crimes by a court where "only those most responsible" were to be held to account, suggests a war where the constitutive parts of strategy were indeed different to those we are used to. Throughout the trial at the Special Court, witnesses from all walks of life testified for and against the accused. Some of the witnesses were themselves Kamajors, others, such as former Vice-President Joe Demby, were involved with the CDF organization from a central position and with the Kamajors in a supportive capacity. Defense witnesses expressed support and admiration for the civil militia and its undertakings whereas prosecution witnesses focused on the brutality of the fighters and the collapse of discipline. But all types of witnesses have expressed again and again the importance of the initiations and the magical protection bestowed by it on the Kamajor fighters. In the face of relentless cross-examination by non-Sierra Leonean prosecutors, witnesses, have maintained a position that accepts and defends the powers of the fighters, derived from the rituals undergone during initiation. In cross-examination by Chief Prosecutor Desmond da Silva on May 15, 2006, a witness repeatedly explained how shotgun bullets would not hit immunized individuals. The prosecutor even brought cartridges into the courtroom to demonstrate the difference between blank and live shots but the witness maintained his stance.[11] Despite the time elapsed since the end of the war (at that time four years) and the application of a primarily Western legal process to extract the truth about events that took place during the war, the belief in the power of magic was undiminished. It would therefore seem presumptuous to assume that these practices were merely instrumental tools, useful during the war but of limited postconflict value. It also suggests that the belief in the power of magic, so firmly shared, would have impacted on how the group devised its strategy and tactics. Looking more closely at the use of beliefs and practices that appeal to a supernatural, spiritual world to wage war is bound to help us rethink the way we think about war as an activity and the place of culturally specific practices in it.

The use of rituals invoking supernatural power to aid and inform a war effort was not unique to Sierra Leone. Insurgent groups in neighboring Liberia, in Nigeria, the Democratic Republic of Congo, Angola, Uganda,

Mozambique, and Zimbabwe have all displayed similar practice on their battlefields. Nor has this widespread phenomenon gone completely unnoticed in academic circles. Anthropologists have studied the interplay of beliefs and practice across the continent for decades and clearly highlighted the relevance of the spiritual world to perceptions of power, both at local and national government levels. The way these beliefs have impacted on behavior during war has been discussed less.[12] Most importantly, the bulk of the discussion that has taken place has done so within the field of anthropology and has not resonated in strategic studies or international relations. This gap needs to be bridged by showing how religious beliefs can be a source of military tactics. In contemporary warfare, ideas and social practices that traditionally fall outside the scope of military studies need to be thought about. After all, the battlefield in these wars is increasingly the population. Winning their hearts and minds requires that we understand their way of thinking. Looking at how we should think about the influence of beliefs on strategic and tactical decisions in an African war can therefore illuminate strategic and tactical problems faced elsewhere—particularly by Western armies engaged in theatres away from their homeland and cultural comfort zone, but equally for any practitioner seeking to influence behavior through nonviolent means to encourage peace.

Organization of the Study

There are several stages to a study of the role played by supernatural beliefs and practices in African wars. We need to first understand what the beliefs we are talking about are and what space they inhabit, how widespread they are and what the pitfalls of their study might be. We also need to think carefully about how best to approach these beliefs so as to not fall into the trap of either oversimplifying or misrepresenting their influence and relative importance.

Chapters 2, 3, and 4 are devoted to briefly disentangling these conceptual issues to guide the subsequent exploration of a specific case study. Chapter 2 is devoted to the concept of "magic." The term itself is heavily value laden, which is in part why it is attractive. Language is one of the worst offenders when it comes to analyzing culturally specific practices and often conveys as much as, if not more than, the analysis itself. But in attempts to clarify and add nuance, the terms that are used to describe the beliefs and practices that alternately get referred to as "magic," "witchcraft," and "traditional religion" can in turn add judgment and new confusion. "Modernity" has in many instances become a measuring stick for the relative contemporary relevance of these practices but making the case for the modernity of magic can add more confusion than it dispels. This chapter argues that although it is important to be aware of the linguistic traps that will inevitably haunt a study of "magic," the labels are less important than the practices themselves, as is judgment on where they fit on a supposed spectrum of the evolution of intellectual thought. Bearing this in mind, supernatural, magic, and religion/traditional religion will all be used throughout this study to convey, in turn, something beyond the natural world and its known laws, practices that are tied to beliefs in the supernatural and intended to proactively influence the material world through supernatural power, and the belief system that incorporates both as a whole.

Chapter 3 looks into the types of beliefs and practices that have made themselves known in contemporary African warfare to draw some tentative conclusions about the roles they play. It shows how the appeal to shared beliefs in a spirit world, the use of rituals and spiritually guided codes of conduct have aided the mobilization and legitimacy of armed groups in conflicts as varied as the Holy Spirit Movements insurgency in northern Uganda and the civil wars in Zimbabwe and Mozambique. Once formed, the practices have also aided groups to maintain internal discipline and helped both to motivate group members and intimidate the enemy and civilian populations. This makes the case for the continued relevance of magic on African battlefields as much as in civilian life and its ability to impact not only individuals—the way religion is generally thought to influence warfare—but the progression of an armed group as a whole.

Chapter 4 builds on these insights to offer a conceptual tool kit that allows us to incorporate spiritual beliefs and practice as an integral part of strategic and tactical behavior. Although strategic theory is reluctant to make space for activity that is not clearly instrumentally rational, borrowing from Pierre Bourdieu's theory of practice we can unpack the interaction of dispositions and resources that shape behavior and find a clear space for beliefs, even those that are anchored in the supernatural. Using this lens we can consider the shared material and nonmaterial context in which these beliefs (and in extension practices) are put to use, the meanings they provide and the functions they serve. This helps us bring new levels of analysis into the study of behavior in warfare, strategy, and tactics that feed into the means-ends relationships the field generally focuses on. Contextualizing the means and ends beyond the immediately military or even political, the inclusion of magical beliefs and practices no longer appear so alien.

Following this conceptual ground work, the case of the Civil Defense Force in Sierra Leone during its 1991–2002 civil war is considered in greater detail chapters 5–8. Looking specifically the Kamajor society, that made up the CDF's most numerous group, these chapters applies the approach identified in chapter 4 to assess the nature and dynamic of their practices.

Chapter 5 lays out the context of the war and the development of the Kamajor civil defense. Chapter 6 then explores the spiritual context and cosmology in the primarily Mende southeastern Sierra Leone from where the Kamajors emerged. Chapter 7 explores the specific role of magic by considering its use in and significance for recruitment and mobilization; whereas chapter 8 looks at the role it played on the battlefield.

Finally, chapter 9 puts these findings from the Kamajor case study in the wider context of strategic thinking and reflects on the implications for the study of war as well as practical engagement in conflict areas. Although the specific case of magic is indeed specific, as is the particular case of Sierra Leone, the logic that informs the formation of practice imbued with spiritual or other culturally colored beliefs is more broadly applicable. At a time when Western militaries and civilians are becoming increasingly involved in conflicts around the globe, the understanding of how specific dynamics impact on warfare will have a direct bearing on the appropriateness and therefore success of strategies—whether employed to secure military victory or nonviolent goals.

Part I

Chapter 2

"Magic" in Contemporary Africa

"Magic"

"Magic" has become shorthand for using invisible supernatural power to affect people and events in the visible and tangible human world. Although in the West it tends to equate with fortune-tellers at funfairs, teenagers dressed in black with pentagrams around their necks, or more sinister groups practicing blood rituals, the contemporary remit of "magic" is in fact much broader. Most mainstream religions include ritualistic practice intended to invoke supernatural powers, be it through prayer, communion, sacrificial offerings, or incantations. Religions and belief systems that have captured a space close to the political power center tend to become more sanitized and their rituals less explicit, whereas belief systems that have not been awarded official status are often more devolved and less doctrinal in addition to maintaining more explicit rituals. "Magic" usually falls into the latter category, even though the type of beliefs behind practices are not fundamentally dissimilar from its more organized and centralized kin in cathedrals, mosques, and temples around the world. For all belief systems where supernatural power (or power not explicable through the laws of nature) plays a central role, the application of that power could be referred to as "magic." While there are important substantive differences between the beliefs of a Vodun practitioner in Benin and a Sunni Muslim in Lebanon, it is important to not separate types of religion too strictly lest we build artificial boundaries that obscure more than they clarify. All supernatural beliefs share some key tenets, notably the ability to think beyond the natural world and adapt behavior to an invisible reality as well as the material.

Nonetheless, "magic" does infer an element of mysticism that holds different connotations than an Anglican mass. It conveys a sense of secrecy and hidden knowledge that can be applied by initiates to affect the world, usually through the manipulation of physical objects. Human agency and the application of that agency to objects in the physical world are central to the practice of magic. These practices place human agency at the center, together with spiritual power, and therefore also awards greater responsibility to the human practitioner—no doubt contributing to the often-negative connotations of these practices. They come to represent the ability of an individual to command spiritual power to ends of their own choosing, rather than an appeal to a deity that may or may not choose to heed the request. It is undoubtedly this element of power that has given magic a negative ring, as it is less easily controlled by centers of power and therefore seen by these centers as more dangerous. As a result, practitioners of herbal medicine in medieval Europe, who could bestow life or death without the involvement of the church, were named "magicians" or "witches" and often persecuted by the state.[1] In the contemporary world, this terminology has stuck and practitioners of similar rituals aimed at the application of supernatural power still go by these names. Although the anthropologist researcher or the local neighbor would undoubtedly be able to see through the terminology to the core of the activity and its social impact, it has helped cement a perception of backwardness amongst observers more far removed. Whether we use magic or witchcraft or indeed mysticism to describe these practices, the essence of the labels all refer to the different, obscure and in the end subversive. This is exactly how the practice of magic is perceived by most in the West, whether the practitioner is a soothsayer in Siberia or a witchdoctor in Ghana. In this sense, the label is as much a reflection of how the practices are viewed from afar as they are a reflection of the nature of the practice itself.

Terminology is bound to be to a significant extent arbitrary, and the key to any label is to explain what lies behind it. In the case of "magic" this is best done through an examination of the practices captured by the label. Although all practices where human agents manipulate supernatural power to an end of their choosing could be labeled magic, in this study we are specifically concerned with Africa and magic in this particular setting. It is a type of practice (quite wide-ranging as we shall see) that is unfamiliar to most political analysts and observers but nonetheless crucial to the everyday life of people in many of the continent's nations. To lay bare the meaning of "magic" in an African context we must therefore look at the specific practices in the African setting.

MAGIC IN AFRICA

"Magic" in an African context resides primarily in the space occupied by "traditional religion." Traditional religion, in turn, captures all religious and supernatural practices that derive from belief systems other than the imported Abrahamic faiths or, in East Africa, Hinduism. In any encyclopedia entry for an African country, the religious affiliation of the population will be given as percentages of people subscribing to Christian denominations, Islam, Hinduism in places and "traditional religions." Although the specifics will differ, most of the "traditional" beliefs center around a local spirit world, mediated through specialist spiritual practitioners that can invoke their wisdom and power through the manipulation of natural objects, to resolve everyday dilemmas. Potions and remedies combine with incantations and rituals to fuse together the spiritual and the natural and exert control over both. As opposed to variants of Christianity and Islam, the common themes of African traditional religions are not so much enshrined in doctrine or even narrative, but rather in nontext-based, yet shared, approaches to the nature of power and man's and the world's relationship to it. Compared to a regular priest or imam, these spiritual practitioners tend to be more proactive in their craft, compelling the spirit world to bow to their will through ritual practice. In this way their practices could be labeled magic as the power to affect change lies as much with the spiritual practitioner as with the spirit world. In other words, whereas a priest or imam would be bound in their interactions with God by a moral code and the inclination of God (as interpreted through defined doctrine), for many of the spiritual practitioners in Africa that draw on local traditional beliefs the options are not determined by anything other than their own ability and skill to manipulate spirit power to the ends they choose.

"Power" is, however, not the realist conception of the coercive capability generally associated with states. Although spiritual power can endow individuals and groups with such coercive power, and has in itself a coercive element, African traditional belief systems approach spiritual power as something that is generally ambivalent and ambiguous. Both the power source and the use to which it is put can therefore be alternately positive and negative depending on the intent of the practitioner who can choose to use the power constructively or destructively.

It is a widespread belief system often existing side by side or even overlapping with Christianity and Islam. Traditional healers or spiritual specialists can be found in almost all cities, towns and villages across the continent, and reference to spiritual phenomena is made in local and national press on a regular basis.[2] It is undoubtedly a belief system that is both alive and well.

However, the belief system is not "pure" in its tradition. In addition to existing alongside other faiths, the traditional beliefs themselves tend to be syncretic blends of one or several of the Abrahamic religions and older local beliefs and practices. As a result, many politicians as well as villagers and urban residents can be found in the mosque on a Friday or church on a

Sunday while also making visits to local healers, medicine men, or spirit mediums. In many cases, the church or mosque also holds a prominent position to the practitioners of traditional religion and become incorporated into their system of beliefs. To many observers this has only cemented the view of contemporary African society as a mirror image of a distant European past where superstition crept into cathedrals and palaces as much as village homes—a conceptual ball and chain that risks imbuing analysis of events where magic plays a role with a bias in favor of explanations that assume backwardness on the part of those employing magic.

Modern Magic? Inhibiting Language

As this bias is widespread amongst certainly Western analysts it is almost inevitable to end up in discussions about the relative "modernity" of magic when arguing for the contemporary relevance of belief systems that incorporate magic as a practice. Modernity has become a measuring stick against which cultural practice is held up to ascertain whether it is in fact on par with practices in the Western world (and therefore rational) or whether it is a regression to a more primitive past (and therefore irrational). The term itself is used liberally and to somewhat different ends by different commentators, but although some refer specifically to the process of modernization (where the modern is an integrated large-scale society characterized by secular and bureaucratic organization) others simply reference the opposite of primitive.

In many ways modernity has taken on connotations of a point of progression and therefore an attribute that must be connected to a subject under study if that subject is to be seen as progressive and, often in extension, rational. This has placed some anthropologists in a bind as the awkward fit of contemporary African society to a Western model of development has led them to attempt to prove the modernity of traditional spiritual practices by advocating an alternative modernity that fits with the African experience.

It is clear that Africa is home to several parallel developments that do not fit neatly into the traditional theories of political and social development that have guided Western intellectuals for the past century. The spiritual dimension to politics is clearly one such dynamic. Like much Western religion, traditional religious beliefs and practices in Africa have proven to be a force available both for reactionary and progressive agendas.[3] Whereas the first postcolonial regimes were eager to eradicate traditional elements from social and political life, the challenges posed both by a failure of many states to provide security (economic and physical) for its population and by an increasingly interconnected world and changing economic realities, have led to an increase in traditional beliefs and practices in all layers of society.[4]

This has happened because tension and conflicts created by a changing social reality have not been successfully addressed by the antitraditional approach embraced by the early postcolonial governments. Consequently people have chosen to consult traditional practices to make sense of the changing world.[5] The prevalence of these as well as other religions suggests they have performed significantly better. Several authors have shown how the wielding of spiritual power through ritual has become integral to the ways in which people respond to the challenges posed by the contemporary world. Peter Geschiere has documented the importance of the proactive use of spiritual power in Cameroon to affect relationships between local communities and the state through the creation of new elites and local-level state-building. There is an impetus for using traditional spiritual practices both to prevent outside interference of the state in local affairs and to draw advantage from changing power relations and dynamics. These practices—be they the services of herbalists, witchdoctors or spirit mediums—offer means of addressing challenges proactively. In this sense it can be used both to level and accumulate power and resources.[6] In a similar way, spirit mediums in Niger have commercialized their services and entered the market economy in a way that is radically opposed to the traditionally charitable role of mediums, who through their possession acquire duties to their societies as laid down by the ancestor speaking through them.[7] This ability to adapt and instrumentalize traditional practices, based on supernatural belief, to address current concerns certainly shows "magic" to be of contemporary relevance.

Some have chosen to see this through the lens of modernity and as reinforcing manifestations of modernity while at the same time being engaged to counter the negative impacts of the same. Although there is much truth in the idea that traditional beliefs and practices (including those involving the supernatural) become tools for both making sense of the world and engaging with it, tying this dynamic to terminology with specifically normative undertones is not necessarily helpful. In fact, such an approach practically requires that traditional practices and beliefs be seen as compatible with "modernity" as a failure to do so would be to equate them with backwardness or a slow progression on the path of socioeconomic and political development. Or, as happens more regularly, it requires a reinvention of the term (in this case "modernity") to accommodate locally anchored nuance. Both of these dilute the value of the term itself and do not necessarily shed a great deal of light on the dynamics of the practices concerned.

Using the traditional-modern dichotomy it is also easy to fall into the trap of pitting one against the other, assuming that if the two cannot coexist one must be using any visible manifestations of the other to its advantage in a one-sided relationship. By this reasoning, if the dominant sphere were assumed to be "modern" it would follow that any traditional practices seen in the political arena would be a conscious ploy by elites to win favor with the less educated masses. Or, in reverse, if the dominant sphere were assumed to be traditional, any seemingly Western-style political behavior would in turn be assumed to be no more than a veneer. Of course, on closer inspection it is very hard to determine to what extent the traditional is adapting to the modern and to what extent the modern is making use of the traditional. The former would imply a preeminence of the nonmodern (in many cases assumed to equate to the backward and misguided) and the latter a preeminence of the modern (in contrast defined by progressive and instrumental behavior).

Both of these approaches lead almost inevitably to normative judgment about the relative progress of the society that hosts traditional belief and practice. Almost as inevitably, such judgment gets in the way of constructive engagement, particularly if the individuals or groups concerned are assumed to be backward and irrational. It would therefore seem best to steer clear of the traps of dichotomous language and try instead to focus on the beliefs and practices themselves, without the distraction of strict categorization. As we cannot do away with words, or their connotations, we will, however, have to adopt some terminology and hope that the qualifications made here will prevent them from being misconstrued. The term "magic" serves a dual purpose as a practical shorthand that captures the proactive manipulation of supernatural power through ritual practice while also connecting with some of the most deeply held prejudices (in the West at least) about behavior as rational or irrational. Looking more closely at "magic" should therefore help unsettle some of these prejudices while at the same time keeping the text more parsimonious. But most importantly magic denotes a *practice*, informed by beliefs in a supernatural world, which invokes power to achieve a desired end without the constraints imposed by strict doctrine or the assumed moral imperative of the source of power itself. The beliefs that inform this practice are of a religious nature, based on African traditions and cosmologies, albeit often infused with elements of Abrahamic religions. Whether or not these practices can be said to be "modern" is irrelevant to this study. It is enough to note that they are clearly contemporary and employed to address contemporary challenges.

FUNCTION OF MAGIC

The mere fact that the use of magic is prevalent in Africa today is enough to warrant a closer inspection, to understand how and why it matters in the contemporary world and, in extension, in warfare. Whether or not the beliefs that underpin it are of equal "stature" to beliefs held elsewhere in the world is neither here nor there both because it assumes a normative judgment on metaphysical matters and because it takes us away from the main objective, which is to understand behavior in order to facilitate more constructive engagement with it. Engaging with magic as a particular cultural manifestation that impacts on behavior, we are therefore better off asking about its functions than its relative merits as a developmental stage of human thought.

Magic, together with the beliefs the practice is based on, can be said to serve three primary functions, all of which are proactive and of contemporary relevance: (1) the beliefs offer an existential service; (2) they explain misfortune; and (3) through the practice of magic they offer a means of redress for perceived ills and can therefore act as a strain gauge or pressure valve for societal tensions. These functions have all had their turn at being favored by anthropology, but they are all still relevant. The functions can be broken down further, and the use to which they are put can vary, but the core remains the same. Let us look at each in turn.

1. Existential Service

The human desire for order and explanations is well serviced by traditions that invoke spiritual power, as are social agendas that can make proactive use of these explanations. African traditional religion, in the same way as other religion, provides an existential service to man. It offers up a conceptual framework that stretches beyond the immediately visible and that bears witness to a greater purpose for existence. Although religion has the benefit of stretching even further, beyond the natural world and science, the same function can be filled by nationalism, philosophy and some political ideology.[8] We should bear this in mind to avoid essentializing the role of religion too far in this particular context as the issue at the heart of the enquiry is the role played by local belief systems in warfare, and these could plausibly be entirely secular yet still fall prey to flawed analysis because of their alien nature. In fact, some interesting comparisons have been made between the role of traditional religion in the Mozambican civil war and that of Communist ideology in other independence struggles in the second half of the twentieth century.[9] It is equally plausible that a comparison could be made between these and the way in which Salafi Sunni Islam has become the guiding worldview for some of the insurgent groups in Algeria and Iraq.[10] Using the word "cosmology" to describe the package that constitutes religious beliefs as well as a locally anchored understanding of the world and one's place in it is perhaps more useful as it conveys the totality of the set of beliefs.[11] All religious beliefs fit within a larger cosmology that guides an individual and group's actions and as such it allows both the individual and group to anchor their existence within a wider whole and provide the meaning that humans crave almost absolutely. A vibrant cosmology becomes a central part of reality, in part through its stated concerns and in part through the norms that develop through and alongside it over time and become part of the common sense shared by those that live within it. As we shall see in the case of Sierra Leone, this shared common sense is only partially consciously constructed and to a great extent springs from a shared set of institutionalized norms, informing behavior and conceptual engagement in more or less conscious ways. Where supernatural belief is part of this common sense, it provides the framework that other assumptions about the world can be pinned to, and a reference point for questions about events. In Africa, explanations are therefore in part provided by traditional belief systems.

2. Explaining Misfortune

This broader existential framework, within which man can make sense of his surroundings and justify means and ends, is closely related to the second function, which is to explain specific misfortune as well as the wider state of affairs in the community, state, world or universe. Like any assistance, the provision of meaning becomes most important in times of despair. As a result, existential guidance—whether through ideology or religion—tends to be more in demand in times of upheaval when people seek explanations for their predicament or sudden changes in the world around them. The need for explanation has driven religious and philosophical thought for millennia and is very likely to continue to do so as long as change challenges the wellbeing of groups and individuals.

As already mentioned, African traditional religion is more agency-driven than the pure Abrahamic religions. Explaining the things (generally material) that cannot be explained by any other means—particularly misfortune—has always been a key function of its discourse. Importantly, these explanations tend to be anchored in the actions of some individual agent. In African communities where traditional beliefs are strong, misfortune is regularly blamed on the use of supernatural power by others in the community. Local news coverage as well as fiction (TV dramas in particular) is peppered with storylines apportioning blame on characters mixed up with witchcraft (which is how magical practice seen to be destructive is usually referred to by those who believe in its power).[12] Misfortune does not happen perchance but as a result of conscious manipulation by an agent. In this sense, the use of supernatural power allows people to answer practically all questions of "why" something has occurred. Importantly, this does not mean that material explanations—abiding by the laws of physics—are not also accepted. But whereas physics explain how something happens, magic or witchcraft explains why. The most widely quoted example of this logic came out of Evans-Pritchard's study of the Azande in the 1930s.[13] Evans-Pritchard observed how the Azande would attribute apparent accidents to malicious intent: Granaries elevated off the ground to keep the grain safe from rats were often used by Azande workers as a resting place as it provided shade during the day. Occasionally termites would eat through the wooden stands of the granary and it would collapse over resting workers. Although the Azande knew the termites were the cause of the collapse (explaining *how* it collapsed), the question of *why* it collapsed at a particular time and over a particular person could only be explained by witchcraft:

> If there had been no witchcraft people would have been sitting under the granary and it would not have fallen on them, or it would have collapsed but the people would not have been sheltering under it at the time. Witchcraft explains the coincidence of these two happenings.[14]

The Azande belief in witchcraft also serviced the need to explain the inexplicable. The idiom of witchcraft continues to address the question of "why" in a similar way for a range of issues in Africa today, for example with regard to Aids. In South Africa, many deaths by Aids are put down to "Isidliso," a disease ("black poison") brought on by witchcraft, by relatives and neighbors. Even when Aids is recognized as the affliction leading to death, witchcraft is blamed for that particular person falling prey to disease. Similar interpretations can be found across southern Africa and have led many communities to think in terms of an epidemic of witchcraft rather than Aids.[15] Further removed from life and death but nonetheless of great significance to many, football, and other organized sports have also fallen prey to witchcraft accusations. In the 2002 African Cup of Nations semi-final in Mali, the Cameroon team coach and his assistant were arrested just before the game for attempting to place a magical charm on the pitch to decide the game in Cameroon's favor. In 2006, the Tanzania Football Federation fined two of the country's top clubs for performing witchcraft rituals before games.[16]

3. Redress and Societal Pressure Valve

Assigning blame not only allows you to answer the question "why" it also opens up avenues for redress. Identifying the source of a problem lets you target it, either to reverse the misfortune or to exact justice. Redress can be constructive or destructive although the boundary between the two becomes blurred by moral imperative. Constructive redress is the resolution of a problem in a way that lets the protagonists (or community) progress without further concern, having righted the wrong that was presenting difficulty. Destructive redress is more akin to revenge, where the problem is not resolved but emotional satisfaction is derived from punishing the culprit for the pain they inflicted. Punishment can of course in many instances be seen as an adequate resolution of a problem and therefore constructive—disincentivizing the culprit from repeating the actions considered harmful or removing them altogether. Although this line is blurred and vigorously debated in most justice systems, in African societies where supernatural power is almost exclusively viewed as ambiguous these lines are constantly in flux. The power that can be used to inflict harm on a rival (e.g., ensuring that the collapse of the granary coincides with certain individuals taking their break in its shade) can also be used to effect positive change (e.g., to stave off thieves from a property or area or allow a local representative to win elected office). In this sense it is not the power itself that is either destructive or constructive but the use someone puts it to. As a result, misfortune can be redressed in a similarly differentiated manner—tapping into supernatural power to punish or to reform. However, in both instances, the power itself offers a source of additional capital to shape the future. Dichotomies like killing and healing or accumulation and leveling, that would at first appear to suggest a negative and a positive, can therefore be turned around depending on circumstances.[17] For example, the use of spirit power to give an ambitious person extra possibilities to enhance his position can be seen by the wider community as a constructive use of such power, while the same power can also be used to curb overly ambitious people.[18] Similarly, the healing provided by spirits in northern Uganda included an act of retaliation against the aggressor that had caused the suffering it had been asked to heal. This retaliation could result in the killing of the aggressor.[19] This ambiguity makes it a malleable tool for addressing personal or communal concerns, although it also leaves action undertaken to seek redress open for challenge by others who may have a different view of what constitutes constructive and destructive agendas. The more organized a practice and the more oversight of it is kept by a recognized body (generally within the community) the more likely it is to be accepted as the suspicion that it is undertaken purely for personal gain will be alleviated.

A particular manifestation of redress that spiritual explanations especially lend themselves to, are accusations of witchcraft that serve to relieve tensions in a community. The "strain-gauge" theory—originally advocated by structural functionalist anthropologists—focuses on how such accusations can serve as a pressure valve for tensions in a society that have no other legitimate outlet.[20] For example, in polygamous marriages neglect, favoritism or differential treatment by the husband can lead to strong tension at the same time as open conflict between wives is socially unacceptable.[21] An accusation of witchcraft can, however, provide an outlet for the resentment, displacing the blame onto the figure of the witch.[22] By accusing an enemy of intending to cause harm, or even blaming them for events that have already occurred, conflicts that are in essence about something different (such as household or business rivalry) can be elevated to a societal concern. As witchcraft is not visible and fundamentally about the use of hidden power, it offers a convenient opportunity for placing blame without the need for a great deal of initial evidence, but therefore also makes events and their causes less predictable and more perilous.[23] The fact that the use of spiritual power by individuals for purposes that serve personal agendas is a cause of suspicion makes it easier to rally community support against it. In the Democratic Republic of Congo, this ease of accusation has been felt acutely by children who have been ostracized and persecuted by communities blaming them for witchcraft. In these children's case, it is often their lack of social connection have made them suspicious to the communities around them. Many have been made orphans by the war or have moved to live with distant family members, either away from the fighting or where prospects for the future are thought to be better. For the orphans there are no parents or families to take responsibility for their actions or instill them with community values and many of the orphaned children are forced to resort to theft and other crime to stay alive in the harsh postconflict environment. Blaming the children for the misfortunes in the war-ravaged country can therefore relieve some of the pressure felt by the community (by identifying the source of misfortune) and also provide a source of redress. But even where there are no obvious scapegoats (individuals or groups alienated from the core community) accusations of malicious practice can be leveled against those less able to defend or explain themselves. In parts of DRC this has led to parents accusing their own children of witchcraft when their family is affected by misfortune. In this case, redress has come through exorcism, expulsion, or death.[24] In a situation where real redress is impossible—that is, reversing the misfortunes brought on by a decade of war and many more of misrule in the DRC—focusing blame on more immediate culprits nonetheless provides a means of remaining proactive and maintaining a sense of engagement with the destiny of the community.

But accusations of witchcraft can also be a tool for manipulating power relations. Accusing individuals in positions of power of practicing witchcraft can undermine their position quickly and force them out of public life. Even if they manage to throw off the accusations and retain their position they are likely to be considered with suspicion, constraining their ability to carry out their office. In Nigeria, top officials of the Niger Delta Development Commission were accused of witchcraft by the Joint Revolutionary Council (a militant group active in the Delta region), effectively dashing any hopes of collaboration between the two for peace.[25] Accusing the body of witchcraft communicated that its officials were thought to be motivated by greed and a ruthless pursuit of advancement through occult means. In this case the accusation played directly into an agenda of political confrontation and was specifically intended to shape negotiations between a state actor and an armed group:

> The recent report of witchcraft and sorcery levelled against top management of the Niger Delta Development Commission has gone further to lend credence to the extent of wickedness, greed and base avarice that continues to drive those who today govern the affairs of the Commission which is supposed to provide social, economic and infrastructural development to the long suffering people of the Niger Delta region.
>
> It has proved beyond all reasonable doubt that the management of the Commission is in the hands of desperate and over ambitious people whose only concern is to enrich themselves by any means possible whether by witchcraft, sorcery, blackmail and threats of assassination.[26]

Witchcraft is mentioned in the same breath as blackmail and assassination, leaving little doubt as to the perceived nature of the practice, in effect to strengthen the accusation as a whole and show the unsuitability of the Commission for the task it has been set. Of course in the same way that spiritual power can be thought to enable practices not approved of by a group or community, it can also be a resource for activity that is perceived as beneficial. This is most obviously the case with healers who provide a positive service for those with ailing health, but also as we shall see through the power given to warriors to improve their ability to fight enemies. Nonetheless, the central importance of interpretation of the motives that drive the spiritual practitioner or his clients makes it a nuanced space where perception plays as great a role as the actual practice.

FROM MAGIC TO WARFARE

Traditional beliefs and the practices that draw on these are still relevant to social and political life in Africa because they are still very much part of the widely shared cosmology of its peoples. They provide means of explaining and engaging with the contemporary world and its challenges by drawing on a register of beliefs that resonate with the wider population. Importantly, these beliefs and practices also tend to be focused on perceptions of power. The nature of power and how to access it are key concerns for anyone engaging in politics, whether in peacetime or during war. In Africa, notions of power are closely linked to ideas of the supernatural.

Social, political, and religious activities therefore become interrelated, if not directly then indirectly through the assumption that all sources of power are tied to the spirit world. In their insightful study of religion in contemporary Africa, Stephen Ellis and Gerrie Ter Haar ague that "it is largely through religious ideas that Africans think about the world today and [-] religious ideas provide them with a means of becoming social and political actors."[27] These religious ideas are by no means exclusively traditional but constitute a space where religious ideas more broadly offer the first port of call for explanation, guidance, and proactive assistance. The prevalence of traditional beliefs must be seen against this backdrop. In this context, power becomes intimately connected to the ability to influence the spirit world, whether through traditional, Christian, Muslim or other practices. Whether national leaders, politicians, or local-level chiefs, all positions of power are generally assumed to be related to spiritual prowess by the population. Poor leadership and good leadership alike is explained with reference to the use to which the leaders have put their spiritual power, as seen in the example from the Niger Delta quoted above. This does not mean that constituents cannot distinguish between nonspiritual corrupt practices or even good governance and a purely religious explanation of events, but as in the case of Aids the explanations tend to merge—spirit power is assumed to give either the impetus for worldly activity or the power to see it through. As an extension of ostensibly political activity, warfare is therefore equally affected by this understanding of power. It makes the spiritual realm a logical resource for activity that requires significant power, whether for a perceived "good" or "ill." As a result, enemy activity is often associated with "evil" use of spiritual power whereas the own side's appeal to the same is considered positive and constructive. Crucially, as we shall see in coming chapters, the shared belief in the power residing in the spiritual world means all action gets interpreted within a similar conceptual framework and regardless of the moral judgment on a particular use of power the fact that it is a power retrieved from the spirit world is much more rarely questioned.

Chapter 3

Spirits in Battle

If magic has been a mainstay of African warfare in this century as well as previous ones it should be reflected in the writings on war in Africa produced in this time. It would be an exaggeration to say that all authors dealing with conflict on the continent reference the use of magic, but many do. In fact some do so extensively and with great insight. Writing about the Holy Spirit Movement in northern Uganda, Heike Behrend shows us in some detail how the group placed spirit guidance and assistance at the core of their military activities.[1] Stephen Ellis was so explicit in his referencing of cannibalism for spiritual enhancement during the civil war in Liberia that he was temporarily sued for libel by Liberia's then president Charles Taylor.[2] Others have touched on important issues relating to the use of magic, if not as exhaustively as Behrend and Ellis. Although most have stopped short of naming function—focusing instead on descriptions of the phenomenon—or using magic as a test case for the wider analysis of culturally imbued behavior in warfare, there are many lessons to be learned from the cases they have studied. For one, scanning the work that does reference magic in the context of warfare, a familiar pattern quickly emerges.

First, fighters are generally offered some kind of protection from the harm of enemy fire and attack. This protection comes in the form of charms, potions, rituals, and codes of conduct. Protective charms have been worn and protective rituals gone through by fighters across the continent. Fighters on all sides wore protective charms in the Liberian civil war, although they were particularly popular with the National Patriotic Front of Liberia (NPFL) during the 1989–1997 phase of the war, and with the Liberians United for Reconciliation and Democracy (LURD) in 2001–2003.[3] Interestingly, LURD in part adopted the practice from some of their recruits from the Kamajor Civil Defense Forces in Sierra Leone who were generally recognized by their abundant use of charms as part of their dress.[4] In Mozambique, Renamo fighters would carry or wear charms provided by their witchdoctors, and in the Democratic Republic of Congo the name of the irregular forces operating in the east, Mai-Mai, refers specifically to the use of magic for which they became notorious in the 1960s ("powerful water" in Swahili). Although in the post-Mobutu war the name has come to encompass a wider range of actors in the east of the country, charms still feature prominently in many of the fighters' apparel.[5]

Similarly, fighters can be imbued with new powers to enhance their ability to fight. Some Renamo commanders allegedly had the ability both to foretell the future and fly, helping them learn about their enemy's plans.[6] Powers to communicate across vast distances or to alter one's physical shape have been seen in northern Uganda as well as Liberia. These powers are often closely tied to the practice of particularly powerful individuals, spirit mediums, or ritual experts. Leaders use the assistance of spiritual practitioners—or acquire these skills for themselves and do away with the service of the middleman—to devise and guide strategy and tactics. Spirit mediums played central roles in Mozambique's civil war and Zimbabwe's war of liberation, as well as in northern Uganda where the rebel movements have drawn their very lifeblood from services rendered by spirit allies. The craft drawn upon to achieve this tends to be the syncretic mix of traditional religious belief and Abrahamic traditions explored in the last chapter.

But in addition to showing us how magic can manifest itself in warfare, the wars in Uganda, Liberia, Mozambique and Zimbabwe also give us a sense of the dynamic this fits into by suggesting a number of functions served by spiritual belief and practice.

Mobilization and Legitimacy

Tapping into the beliefs of the local population and securing the support of key spiritual leaders can aid in the mobilization and recruitment effort of a group. In this sense, a spiritually referenced framework can allow the cause and aim of the conflict to be framed. This pattern has taken slightly different shape in different settings, but a general trend can nonetheless be identified.

In Uganda, the spiritual rhetoric was the very raison d'être of the Holy Spirit Movements (HSM).[7] For the HSM and its armed wing the Holy Spirit Mobile Force (HSMF) the end in sight was a spiritually cleansed and healed Acholi people able to move on from the sins of its past that were hindering development and progress.[8] These sins were the killings during the previous civil war (between the National Resistance Movement (NRA) and the Uganda National Liberation Army (UNLA) in 1981–1986) in Luwero, and the failure of soldiers to respect traditions upon their return, leading to haunting by *cen* (the vengeful spirits of the dead) and rifts in Acholi society.[9] Alice Auma, the medium of Lakwena (more commonly known as Alice Lakwena), organized the HSMF to wage war against the government and witches to redress the misfortunes brought on by this sin.[10] The rhetoric built upon the spiritual tradition in northern Uganda that recognized spirits as responsible for misfortune and catastrophe, but also as a power against disaster. The disruptive civil war between the NRA and UNLA, and the ensuing social unrest with returning soldiers, were interpreted by Alice Lakwena as a spiritual matter in need of spiritual remedy. This rang true to large parts of the population that had already interpreted misfortune and death in war, or resulting from war, as a manifestation of witchcraft for some time.[11]

Although the campaign waged by the HSMF can be understood also as a response to socioeconomic marginalization and crisis, these factors were *framed* within a spiritual understanding of the causes of these grievances.

For the HSMF's successor movement, the Lord's Resistance Army (LRA), establishing a society based on the biblical Ten Commandments was the initially stated aim of the struggle. However, this aim has been expressed much less clearly and less consistently than that of the HSMF. For example, in the political wing's manifesto the emphasis was on human and economic rights and less traditional or religious in its language, and any stated political agenda has been hard to come by. Even those with the most access to the LRA high command are reluctant to articulate any agenda on the part of the rebels, nor have those participating in the Juba peace process put forward any consistent vision for the future.[12]

Spiritual motivation for the war does, however, manifest itself on other levels. The most interesting is perhaps the dispute over a blessing that was supposedly given to the LRA in the early days of their struggle, and the curse that is supposed to have been cast on them subsequently: the LRA leader, Joseph Kony, claims he received the blessing of Acholi elders to go and fight President Yoweri Museveni's forces. Blessings of this sort are often given in times of warfare and come in a number of forms. After presenting a valid reason for warfare, accepted unanimously by the elders, the leader of the war party may be blessed in front of the ancestral shrine and presented with *lapii*—sticks for making fire symbolizing legitimate authority.[13] A more general blessing could be given by presenting the fighters with branches of the *oboke olwedo* tree, and individual elders could bless their sons separately to protect them in battle.[14] Once given, such a blessing could not easily be retracted. That Kony and the LRA were blessed in this way is widely disputed, and even if he did receive some type of blessing, most believe it is unlikely he was presented with *lapii*, that is, given the legitimate authority to wage war.[15]

Nonetheless, part of the justification given by the LRA for their continued struggle against the government *and* their turn against the civilian population in Acholiland, has been the blessing and mission given to them, and the subsequent betrayal of this blessing leaving them abandoned, and in part bound through the blessing, to a struggle no longer supported by those who inspired it. The betrayal is seen both in the denial of the blessing, but also in a curse supposedly cast on the LRA.[16] In this sense, the struggle remains framed within a spiritual context that both explains the origins as legitimate (the blessing following the elders consensus that the cause was just) and the course of the war as tied to the overturning of the blessing once given. However, within this frame the more immediate concerns have no doubt become survival and a fear of repercussions should the group return to civilian life.

Spiritual claims were central to the mobilization of support, particularly for the HSMF. It lent legitimacy to the movement and promised greater chance of success with the aid of spiritual power. For the LRA the role of the disputed blessing highlights the importance of spiritual sanction for the legitimacy of, and subsequent support and mobilization for, the force. In a way the denial of the blessing can be seen as an attempt to undermine this mobilization tool by questioning the legitimacy of the cause—not unlike propaganda wars fought alongside the physical battles in wars across the world. Here, the spiritual is a central part of the cultural context, and therefore becomes central also to the psychology of the war and its actors.

However, the mobilization and recruitment of fighters to the LRA has been largely forced through the abduction of civilians and primarily children. This stands in stark contrast to the HSMF where fighters joined the force voluntarily. Why the LRA have not attempted a similar mobilization strategy as that employed by the HSMF is unclear, although they are likely to have faced the same difficulties as did the first of the HSM successor movements, Severino Lukoya's Lord's Army. Severino Lukoya, the father of Alice Lakwena, took over the struggle after her defeat. But he had trouble redefining the struggle that had just failed so dramatically with the disbandment of the HSMF after a crushing defeat, in a way that could convince people to once again take up arms or support the fighters. The campaigns of the HSMF had been draining on the north, emotionally as well as physically, and the movement had become increasingly harsh in its dealings with uncooperative civilians. As Severino entered the scene he was therefore faced with a more suspicious population than Alice had been a couple of years earlier, and never managed to secure enough support to engage the NRA in any sustainable way. The LRA, which emerged around the same time as Lukoya's group is therefore likely to have faced the same suspicion and fatigue, and as a result forced to consider alternative means of building up its fighting force.

The motivation for soldiers to join the movements was boosted not only by spiritually enforced legitimacy but also by the promise of the support of spiritual power. Immunizations against enemy fire made battle less intimidating. In addition to this advantage, the troops were promised direct spiritual reinforcement. According to Behrend, the fighters of the HSMF, once killed in battle, would stop over briefly in purgatory before returning to the battlefield as spirits to fight alongside the living. At any one point in time, the HSMF was supposedly supported by 140,000 spirits in battle.[17] In this way, the promise of the afterlife provided yet another incentive to join the ranks. This is not unlike the rewards envisaged by many Muslim suicide bombers in the Middle East and Asia, where promises of a privileged afterlife are also offered as incentives for self-sacrifice. But more immediate rewards were also promised to the fighters; Behrend cites a reward of 15 children after victory, and another of a car and pretty house.[18]

Similar experiences are in evidence elsewhere. In Zimbabwe, the Zimbabwe African National Liberation Army (ZANLA) created close links with spirit mediums at the early stages of the war of liberation.[19] ZANLA's first encounters with local mediums allegedly led to an alliance only once it became clear that such a liaison would help legitimize ZANLA and its cause in the eyes of the local population. Most of the ZANLA soldiers were exiles and had grown up outside traditional Zimbabwean society, and had developed a secular socialist ideology, which also partly fuelled their struggle. However, the power of the spirit mediums was quickly recognized and the recruitment strategy shifted to incorporate the mediums. Lan recounts how when ZANLA recruiters first met with local communities upon entering Zimbabwe they were told they had to speak to the elders who in turn told them to consult with the mediums.[20] Instead of going into villages and recruiting young people directly, the rebels would therefore approach the mediums and with their blessing hold rallies explaining their ideology and cause to prospective fighters. Local resentment of the collapse of traditional land rights and hierarchical structure under Rhodesian rule made it easier for the rebels to gain the acceptance of the mediums, as did the fact that there was a spiritual order into which the fighters could be placed as warriors on a par with some of the most prominent ancestral spirits. The spirit mediums had assumed the political authority previously exercised by the traditional chiefs who had become too bound up with the Rhodesian state to maintain their legitimacy in the eyes of the population. The mediums that represented the spirits of deceased chiefs (mhondoro), represented the authority and legitimacy of the respected chiefs of the past. This legitimacy they passed on to ZANLA guerrillas by naming them the successors of the mhondoro.

In Mozambique, the Resistência Nacional Moçambicana (Renamo), introduced a spiritual rhetoric to draw on the grievances of the rural population dissatisfied with the government's, and later Frelimo's (Frente de Libertaçâo de Moçambique), dismissal of religion. In fact, Renamo played on all religions represented in Mozambique and its diasporas to broaden their support base as much as possible. Although some have suggested that they chose to adopt a "spiritual" agenda only after they had read the work of academics David Lan and Terence Ranger and assessed the success of ZANLA's campaign in Zimbabwe, it is important to look at how this agenda was played out.[21] And indeed, it is testament to the strategic value of creating links with spirit mediums that Renamo chose to go down that road, regardless of the exact causal nature of the original decision. As such, it certainly highlights the potential for external manipulation and a more immediately instrumentalist approach to the adoption of spiritual practices. But the recognition of the utility of relating to and drawing on local structures and practices must be seen in a wider context where those practices make sense and reflect dispositions and experience that are generally shared—it becomes a corner stone of strategies aimed at winning the support of the population, either through genuine support or fear.

In Mozambique the traditional religions were the most potent.[22] Renamo therefore based their mobilizing rhetoric on an appeal to the revival of practices denounced by Frelimo, such as those of feiticeiros (spirit mediums) and curandeiros (traditional healers).[23] But Renamo also adopted a more aggressive spiritual rhetoric, for example blaming the drought of the early 1980s on Frelimo's alienation of the ancestors, and claiming that their struggle was "of the spirits," or "for the ancestors."[24] For both ZANLA and Renamo the traditional tribal system of governance proved a good framework for administering local populations once the institutions of power such as the mediums had been anchored in the rebel camp. This was often accomplished through the recognition of the ills traditional religion had suffered at the hands of the respective enemy.

The potency of the spiritual language became truly evident, however, only with the emergence of a countermovement—the Naprama. The Naprama used the same spiritual language as Renamo but with an even stronger focus on reviving traditional religious institutions. Although Renamo had initiated the rhetorical dialogue of a war of the spirits, others were not late in following up, and expanding the subject matter beyond the interests of Renamo and for their own purposes. Frelimo did the same as it became obvious that the spiritual dimension was playing into the hands of Renamo. Instead of attempting to expose Renamo's activity as fraud, Frelimo attempted to prove their own credentials in the spiritual world by linking up with local spirit mediums and adopting a spiritual rhetoric.[25] K.B. Wilson retells one such story of how Frelimo managed to capture one of the important mediums of the Renamo camp and displayed him around a government-held town as a morale booster. The capture was also described in spiritual terms, one Frelimo soldier supposedly going into the Renamo camp protected by magic and abducting the medium unseen.[26] The adoption of this type of rhetoric by all sides escalated the spiritual dimension of the struggle. In the early 1990s, Renamo officials spoke of challenging Naprama to prove that the ancestors who supported Renamo were more powerful than those who favored the Naprama leader Manuel Antonio.[27] Naprama spoke of opposing Renamo to allow for the true reinstatement of tradition, and Frelimo used its own magic to catch up with both.

In a similar manner the Rhodesian government took on board the spiritual language of the mediums once its potency had been proved. Propaganda leaflets would be dropped from airplanes with appeals from prominent spirit mediums to stop aiding the rebels. In one such leaflet titled "A message from the great medium of Mhondoro Mutota to all his people," a medium helping ZANLA was accused of violating the ancestors and denounced as a criminal. It went onto state that "the ancestral spirits do not want blood-shed in this country. I shall work with the government. I shall not die helping terrorists."[28] This stands in stark contrast to earlier statements made about the rebel movement by Prime Minister Ian Smith where he concluded that "they have found a few witchdoctors of doubtful character and of little substance, and succeeded in bribing them to their side... I am sure I do not have to inform you how easy it is to mislead these gullible people who still believe in witchcraft and the throwing of bones."[29]

Organization and Discipline

The spiritual also comes into play within the ranks of the respective movements to facilitate organization and discipline. The use of immunizing rituals is perhaps the most evident of these strategies, and returns in settings across the continent. The rituals and medicines differ but the outcome is generally the same: bullets turn into water or miss their targets; soldiers become invincible and occasionally acquire supernatural powers such as flying or making themselves invisible. The two latter powers tend to be reserved for the higher ranks and form part of the organizational structure. Their extra powers can then be used as an instrument in establishing hierarchy as well as in maintaining it, and are reinforced through the retelling of missions in a language filled with reference to these powers.

In the successive Holy Spirit Movements the leader was primarily responsible for spirit communication. In Alice Lakwena's HSMF, Alice had a monopoly on possession and conferred messages from all members of the spiritual board that ran her movement. According to Behrend, the spiritual hierarchy of the HSMF consisted of a number of spirits, each of which had a different area of responsibility, headed by the "chairman" Lakwena. For example, the spirit Wrong Element was in charge of intelligence, Franko for food supplies, and Ching Poh for weapons supply and transport.[30] In this way Alice effectively controlled all aspects of her movement through the spiritual head of each division. This monopoly was broken in the movements that followed the defeat of the HSMF. Her father's Lord's Army allowed anyone to be possessed by a spirit, although such possession was seen as a duty to take up arms and kill. In the latest and current incarnation, the LRA, the proliferation of spirits has been even greater. Although Kony maintained a monopoly of possession by the most important spirits, he has let other mediums join his movement and have their own spirits.[31] As opposed to movements where advancement meant spiritual favor, the Holy Spirit movements in northern Uganda have ensured discipline through the practically almighty power of the leader through his or her connection to the spirits. The linking of possession and duty to fight also emphasizes the greater responsibility of those in touch with the spiritual dimension, and indirectly establishes a hierarchy.

The Holy Spirit Safety Precautions were another mechanism for ensuring discipline. Alice issued 20 rules to guide the behavior of fighters, in camp and on the battlefield. Rule 14; "thou shalt follow the right words of command, and never argue with the commander" did so explicitly. But the absolute power of Lakwena lent absoluteness also to the rules. Rules 18, "thou shalt not branch off to any home or shake hands with anybody while on route to the battlefield"; twelve, "thou shalt not pick from the battle field any article not recommended by the Lakwena"; and nine, "you will execute the orders and only the orders of the Lakwena," were also directly disciplinary in nature.[32] Failure to comply with the rules would remove spiritual protection from a fighter and invite death from enemy fire.

The LRA may have done away with the Holy Spirit Safety Precautions in favor of an adoption of the Ten Commandments as they stand, but use initiation rituals similar to those of the HSMF and retain similar disciplinary rules. Newcomers get sprinkled with holy water and their bodies smeared with shea-butter oil and white ashes. In addition to the Ten Commandments they are also forbidden to drink or smoke or eat certain types of food—just like the HSMF—and the breaking of these rules result in severe punishment. Abductees have also told of rituals aimed more explicitly at the intimidation of recruits for purposes of discipline. Initiation involves a beating to introduce them to life as soldiers, sometimes with the killing of one or more of the recruits, followed by a ritual binding them to the LRA. Shea nut oil was then smeared over each recruit's forehead, chest, back, hands, and feet in the sign of the cross and they were told "the shea-nut oil would make it easier for the LRA to find them should they try to escape"—implicitly suggesting that the ritual would give the commanders spiritual power over the initiate, including to track him down.[33]

Whereas the elements of intimidation and explicit discipline appear to be primary in the LRA, the rituals bear significant resemblance to those of the Kamajor initiation practice in Sierra Leone (see chapter 7), particularly the bonding together of fighters through the shared experience of initiation. However, the forced nature of LRA recruitment (and as a result initiation) is likely to have made it weaker, as evidenced by the high degree of defection and escape by fighters when given the opportunity.

In Mozambique, Renamo missions, successful thanks to good intelligence, were often recounted as the result of the commander's ability to read the future and anticipate Frelimo movement. In Renamo, the different levels of power and spiritual initiation were further reinforced through a differentiation of ritual practice. Some ex-combatants describe how a witchdoctor would mix goat's blood with herbal medicine for fighters to drink before battle to prevent them from dying in battle. The commanders on the other hand were recommended to bring the medium the inner organs of newborn children so that she could make them more powerful drugs.[34] Within Renamo, the elite Grupo Limpa, were tasked with conducting key operations, but also with enforcing internal discipline.[35] Knowing that the higher echelons of the movement have access to the more powerful rituals and the spiritual protection and power they endow, acts to reinforce hierarchy and adds a potent dimension to that of mere rank insignia.

Initiation into the forces of a movement drew on these spiritual regalia to entrench the sense of a supernatural order and hierarchy with the new recruits. The idea of the greater potency of human sacrifice and the exclusivity of the benefits to be gained through such rituals informed rituals by fighting groups in Liberia. Drawing on familiar traditions and structures, but redesigning them for the context of war and the particular group, part of the NPFL training consisted of a ritual distilled from traditional initiation ceremonies. Fighters were tattooed or received small cuts to become immune to bullets. More interestingly, however, a gun containing blanks was occasionally fired at new recruits to prove the success of the rituals.[36] This is not to say that the higher ranks did not themselves believe in supernatural power, even though the "real" rituals may have been withheld from new recruits—a system which bears witness to an additional hierarchical level, although one more deviously cloaked. The same practice was present amongst the Kamajors in Sierra Leone, often with live bullets suggesting a greater belief in the actual potency of the power.[37] We will look in greater detail at this dynamic in chapters 7 and 8. The sense of being part of a spiritual order was nonetheless impressed on the fighters that could then more easily be disciplined with reference to the supernatural. The behavioral codes issued to fighters to ensure continued protection was perhaps the most immediate form of discipline related to spiritual sanctions.

With prospects like the acquisition of spiritual power, adhering to the order promoted by the mediums or ritual specialists (either directly or through military leaders drawing on their expertise) is attractive. But perhaps more importantly the prospects for those failing to obey the rules of the spirits were powerful incentives to stay in line. Alice Lakwena actively pursued impure soldiers and witches not abiding by the rules set out, sometimes resulting in executions, and in Mozambique, Renamo leader Dhlakama was known to have told soldiers they would be pursued and killed by the spirits of Renamo if they defected.[38]

Motivation and Intimidation

A third set of functions represents two sides to the same coin—the motivation of one side's fighters and the intimidation of fighters on the opposing side. Motivating one's own fighters to go into battle and intimidating the enemy (and keeping the local population in check) through spiritual practice and rhetoric has proved successful to a varying degree.

In Uganda, Behrend suggests that spiritual intimidation played into the hands of the HSMF primarily in the early stages of their struggle. She concludes that the shared beliefs of the HSMF and enemy soldiers meant the latter often ran away when they heard the Holy Spirit soldiers singing hymns—a sign of their spiritual endowment.[39] However, as the brutality of the HSMF against civilians grew, support weakened, and the government forces could regain the military upper hand. Spiritual power seems to also play a part in the intimidation of civilians in northern Uganda today, where many believe in the potency of Kony's power even if they do not believe in its legitimacy. Testimony to this effect have been given by returnee ex-combatants and was also seen in an offer to the government from the national council for traditional healers and herbalists association (Nacotha) in 2003; the association offered to assist the government by providing spiritual support for the army against the LRA to facilitate victory.[40] However, it would seem that in present day northern Uganda intimidation of the civilian population is achieved primarily through violence, killings, and the abduction of children.

Examples of a more active engagement of spiritual reference and practice for purposes of intimidation emerge from the other three wars. Creating myths of invincibility is perhaps the most potent of both intimidation strategies and combat motivation. One example quoted by several authors is the André Matsangaisse myth in Mozambique.[41] Matsangaisse was the first military leader of Renamo and had reportedly acquired magical powers from a local spirit medium. Because of his spiritual protection he supposedly tricked death on more than one occasion, and died only when he was betrayed by a spirit medium that had been alienated by Renamo.[42] A variation of this story is retold by Alex Vines, in which some of Matsangaisse's men had broken the rules of conduct required for spiritual protection, in this case respecting the local population, and hence lost their immunity. Unaware of this breach on the part of his men, Matsangaisse went into battle unprotected. The myth is one both of intimidation and motivation in its tale of invincibility, but also one of discipline as it portrays the dangers of departing from the laws laid down by the spirits.

But Renamo pursued myths of invincibility in more ways than the retelling of the fate of Matsangaisse. They generally went to great lengths to remove their dead and wounded from the battlefield to convince their opponents they could not be defeated.[43] If a Renamo soldier was nevertheless killed, medication distributed by a medium or healer before battle ensured the spirit of the dead would be resurrected with the ability to retaliate against those responsible for his death. And Renamo made sure to exploit tales like these to their full extent when describing their missions and progress. Intimidation of Frelimo soldiers thanks to such rhetoric and practice went a long way to demoralize the enemy troops, and played well into the hands of Renamo.

Similarly, the macabre video of Liberian President Samuel Doe's torture and execution that was circulated in Liberia and elsewhere likely served a dual purpose: On the one hand it was certainly a show of force on behalf of the responsible Prince Johnson's Independent National Patriotic Front of Liberia (INPFL). But it has also been argued that it provided "proof" of the President's death, despite his spiritual protection through rituals and charms, and hence, evidence of the superior power of his killers.[44] In much the same way as Renamo explained their victories in terms of spiritual superiority, responsibility for the death of a man who was known for his heavy reliance on religious ritual and links with powerful mediums and priests spelled advantage on the spiritual battlefield in Liberia. Their success was seen as a sign of power, and the proof of success (the video) could establish the source of power in the minds of those it reached.

Drawing on spiritual imagery for purposes of intimidation was also evident in Liberia and the DRC through cannibalism. In Liberia reports have emerged of ritual killings and cannibalism on all sides of the conflict.[45] The reason Liberian President Charles Taylor attempted to sue academic Stephen Ellis for libel was that Ellis had written about his membership of a secret elite society known as the "Top Twenty" practicing cannibalism.[46] LURD, one of the insurgent groups fighting the Taylor government in the last round of hostilities, were also rumored to be practicing cannibalism on the battlefield. In a documentary following the rebels on their campaign, LURD fighters proclaimed that "We eat human heart to be strong." This statement builds on a heritage of cannibalistic imagery and practice that have emerged from the conflict throughout the 1990s. Interestingly, these fighters were Kamajors from Sierra Leone, showing the cross-border relevance of the beliefs these practices draw on.[47]

The practice serves a dual purpose: in part feeding the belief of the fighter that eating the heart of a strong man will transfer some of that strength to the eater, and in part intimidating bystanders either into flight or submission. Civilians and journalists have described how the "choice method of killing" by ULIMO factions in 1995 was to tear into a person's chest and extract the heart while the person was still alive. Also then fighters were encountered that boasted of having eaten human hearts in the belief that it gave them strength.[48]

In the DRC, the Mai-Mai would cannibalize enemies in a similar way to fighters in Liberia. They would particularly target pygmies, some argue because they were assumed to be particularly knowledgeable—a knowledge the Mai-Mai would come to own through the ingestion of their bodies.[49] Locals in Manono, Katanga province reported how a particular group of Mai-Mai, known by the name of its leader Colonel Gedeon, would "wear amulets of hands and fingers, and when they are fighting they nibble pieces of flesh to intimidate their enemy."[50]

The instrumental agendas of ritual killings and pre-battle ritual could plausibly be compared to other motivating forces and tools for intimidation and for example ethnic cleansing elsewhere, such as extreme nationalism. Intimidation is most effective when it connects with the register of beliefs held by the opponent, particularly as they relate to perceptions of power and the ability to inflict pain or death. As the cosmology of traditional religion centers largely on power—whether used for constructive or destructive purposes—it is eminently suitable as a framework for intimidation of this kind. But the use of traditional religious practices and beliefs as a source of inspiration for tactics aimed at intimidation is only culturally specific in its point of reference as opposed to in its inherent strategy. In contexts where ingestions holds connotations of power, cannibalism will undoubtedly be more piercing as an intimidation strategy than where it does not. Similarly in contexts where virginal purity is essential to social interactions, rape will appeal not only to the horror of the act itself but to the undermining of a social code, becoming an attack not only on an individual but on a community.

DIGGING DEEPER

In the wars we have touched on above, magic, and the beliefs that underpinned it, served several clear functions. It helped provide legitimacy, assist mobilization, maintain internal discipline and intimidate the enemy. This is undoubtedly a useful insight on its own, and goes some way toward illustrating how even a scratch at the surface can reveal purpose behind practices that might otherwise suggest chaos and primitive brutality. But identifying these apparent functions also gives rise to some questions. The most significant question relates to the nature of these functions—are they purely instrumental, manipulative or something more? And, more importantly, what are the implications of adopting a functional view of the use of magic as part of an analytical framework tackling contemporary African conflict?

We return to the question of rationality, and whether the functional dimension to magic or religious belief is merely a tool employed cynically to manipulate actors and events, or a consideration informed by spiritual belief also in the higher echelons of a fighting group? It is impossible to get beyond this question without considering in greater detail the specific motivation of the actors involved. Although these beliefs and practices clearly serve instrumental purposes for the implementation of tactics and discipline, there is another dimension that suggests there is more to these than instrumentality. The case of the HSMF is perhaps the most striking. In this case soldiers went into battle not only with limited arms but also with instructions—at least in the early stages of their campaign—not to kill or even aim a weapon at the enemy. In addition to relying on spiritual protection for the safety of the soldiers, the camp was left unguarded. Instead of guards, "the technicians, the ritual experts, blocked the routes to the camp with magic, so that no enemy could approach."[51] This suggests a belief so strong as to ignore safety precautions of a human kind completely, something a nonbelieving and calculating commander, only interested in motivating the troops would be reticent to do, unless he was an avid and reckless gambler. It also suggests that there is something at play that is tactically unsound from a traditional military means-ends perspective. This something is derived from the cultural context of the agents involved, and indicates a distinctive space for local context in shaping tactical decisions.

Having seen how belief in the potency of the spiritual dimension is prevalent across contemporary Africa, it would seem remiss to assume that in warfare it is used as a tool by nonbelieving elites seeking to manipulate the less enlightened. Without a core of real belief amongst both fighters and the population, drawing on these beliefs for instrumental purposes would be nonsensical. And if this is the case, it seems a bit presumptuous to assume that the leadership are completely separated from the worldview shared by their societies. Whatever the cause, if it can be anchored within an existential framework or system of meanings already subscribed to by a population, it gains a significant advantage in terms of all the functions identified. But this requires real belief and although cynical manipulation of such belief is indeed possible, it is not a given outcome. Extrapolating from this it appears logical to assume that magic plays a *real* role and is more than just theatrics. It also seems appropriate to assume that to understand this role we must understand the cosmological, cultural and indeed political context in which it becomes meaningful, to then be able to trace the connection between practice and purpose and understand its impact.

CHAPTER 4

THINKING ABOUT PRACTICE IN WARFARE

Evaluating the role of spiritual practice in warfare throws up a number of challenges. For a start, even though we can define magic as a type of practice with contemporary relevance, as seen in chapter 2, and identify examples of magical practice on African battlefields as in chapter 3, to reach deeper below the surface we need to look more carefully at specific practices. If we do not, our analysis risks straying into the territory of cultural essentialism, putting the use of magic down to the simple fact that "it is their culture" so it makes sense to them. Although this is not technically false, it does not help us understand why particular behavior is deemed appropriate at particular times, nor the extent of its impact on the war as a whole. It also does not allow us to incorporate the consideration of behavior like the use of magic into strategic theory, which is necessary if we want to include the analysis of cultural factors systematically in our study of warfare. To do that, however, we have to first come to terms with some apparently contradictory sets of principles.

Religion is generally thought of as a primarily metaphysical activity whereas warfare is seen as primarily material. Material and metaphysical elements blend in both cases, as they are both realms involving the material world as well as human beings with all their metaphysical considerations. Nonetheless, the main points of reference for the respective realms are opposites: for religious or spiritual belief it is God or the spiritual world and man's relation to it; for war it is the physical engagement of opposing material forces for the purpose of forcing the other side into submission. In addition to this intrinsic difference, the study of religion continues to be haunted by the wider image—certainly among academics, but also popular in the increasingly secular West—of religion as something irrational. Drawing on an Enlightenment heritage, where magic and religion are at the bottom of the ladder of progress topped by scientific enquiry, religious beliefs and practices have tended to be seen as signs of primitive intellectual development. As a result it has until recently tended to be excluded from serious consideration in analysis of "rational" activity such as military strategy.[1]

Strategy is perhaps the most "rational" of wartime activities.[2] The traditional strategic theory that strategy and tactics in practice are supposed to rest on, assumes as underlying principles instrumentality and rationality. Undoubtedly, magic and supernatural force may at first seem incompatible with such assumptions. But as with most existential frameworks that provide explanation and meaning to individuals and groups, the objective "truth" of the beliefs is not what determines the functionality of the actions that result from them. It is with reference to these basic means-ends relationships—which lie at the heart of strategic thinking—that religious beliefs and practices must be related to evaluate their role in warfare. As a result, to understand the behavior of fighters in places where "magic" is a recurring feature one must understand its rational and instrumental purpose—its function.

But while looking for function it is equally important to not fall into a simplistic functional*ism*. Seeing religious ritual and practice solely as objectively functional activity risks misunderstanding the nature of the functions they do serve, which stem in part from an unconscious application of principles and dispositions that emerge as natural and common sense from the worldview subscribed to in a given context. That is to say, the functions can only be truly appreciated (and in extension evaluated with reference to strategic, tactical or other practical relevance) if seen not in isolation from, but as an active constitutive part of a wider worldview and shared material context.

The framework proposed by French sociologist Pierre Bourdieu for approaching social practice is a very useful tool for this exercise. Bourdieu's theory of practice maps the formative and responsive environment of an individual and society and the creation of practice in it. It allows us to bring out the logic of practice that may at first appear neither rational nor functional. It provides a framework for thinking about all forms of practice and its genesis, and as such it applies to strategic *practice* (in the military sense) even if it is perhaps less compatible with certain elements of strategic *theory*. But even in relation to strategic theory it does not so much stand in opposition to its basic tenets as it broadens and deepens the meaning and remit of the constituent parts of such theory. In other words, the underlying principle of rationality that runs through strategic theory, seen through a Bourdieuan lens becomes not just a question of objectively maximizing outcomes by choice of means, but one of unpacking why certain means are thought to maximize outcomes in a given context, and how that context can allow the means to function in a way in which they would not necessarily function outside of that context. The means employed in warfare therefore take their specific form in part as a result of the material context and the options available to the agents operating in it, but in part also as a result of the interaction of the dispositions and resources of these agents in the particular setting. In this way religion or spiritual belief, and the practices that derive from it, come to shape the form of war and the behavior of fighters.

Leaning on Bourdieu we can extrapolate a framework for thinking about the continued relevance of magic and its role in warfare that can accommodate both the material and spiritual, without succumbing to either purely functionalist or relativistic explanations. Borrowing Bourdieu's approach to the logic of practice we can think more usefully about the functions of ritual and magic, without falling prey to a functionalism, or even structuralism that strips the agents and their world of agency. We can then explore the areas where religious or spiritual belief can have an effect in war and where it fits in with strategy.

Rationality, Function, and Logic in Warfare

When considering warfare, strategy and strategic approaches are natural points of reference, if not starting points, for analysis. When exploring the role of spiritual beliefs and practices, as mentioned above, the nature of strategy can seem at odds with the nature of religion or spiritual belief. This type of enquiry can therefore often be relegated to the realm of psychology or rewritten as a cynical and instrumental manipulation of the psychological needs of fighters for the material purpose of the rational strategy at hand. We want to avoid doing either. We therefore need to consider the underlying assumptions of traditional approaches to strategy and tactics with specific reference to the notions of "rationality" and "logic" that would appear to exclude any influence from spiritual beliefs. The intention is not so much to challenge the basic tenets of strategic theory or strategic approaches, but to look more closely at how these can in fact allow for the inclusion of what may at first appear irrational. Bourdieu helps us do this through his theory of practice, but before we turn to that, and to make the connection clearer, the role of rationality and function in strategy needs some more clarification.

Strategy is in part defined by rationality—the rational process of designing and pursuing action for given ends. In addition to relying on an instrumentally rational approach to decision making, and the means-ends relationships that define it, on a theoretical level strategy traditionally also assumes moral neutrality and an understanding of power relations closely tied to the Realist paradigm (defining power as the coercive capability generally associated with states). Means-ends relationships therefore become central to strategy and, more operationally, tactics. Instrumental rationality is the process of reaching an end through the "right" means, generally understood as those that are most effective and efficient. We can therefore conceive of rationality as an internal logic in a means-ends relationship, distinct from normative judgment or morality—or, "the choice of the right means to an end that you wish to achieve... [that] has nothing whatever to do with the choice of ends."[3]

This is the approach to rationality traditionally favored by strategic theorists. One of the key contributors to contemporary strategic thinking, John Garnett, acknowledged its limitations in that

> theoretically rational behaviour involves more than that. It involves the selection, from among alternative possible actions, of that which from the point of view of an omniscient and objective observer, is most likely to promote the value pursued.[4]

However, he recognized that in terms of strategy and violent aggression where the nature of ends and means are as a rule disputed by the opponents, the main concern is with the rationality of the *process* in the means-ends relationship, not the absolute rationality of the choice of means or the end itself:

> in the context of strategic analysis, rational behaviour refers to behaviour in which the actors *try* to maximise their value positions, as well as to behaviour in which they *do* maximise their value positions.[5]

In other words, to understand the adoption of particular strategies or tactics, and to see them as strategic or tactical, we must identify the logic of the means-ends relationship as conceived by the actor(s) in question. We must explain why a particular action is thought to maximize the position of the actors—even if it does not always do so—and why a practice is created or chosen as a result of this; in other words, the *logic* of the means employed in this instrumentally rational means-ends relationship embodied by particular strategies or tactics.

In his book on the strategy of the Irish Republican Movement, M.L.R. Smith suggests an approach for evaluating the strategy of an actor, the essence of which is

> simply to trace the line of thinking of a particular political entity in order to comprehend how it proposes to achieve its objectives; and also to *look at the ideological assumptions and values that underlie that entity's thinking and how this informs the way it formulates strategy.*[6]

The necessity of such an approach applies also to the level of tactics and more generally to battlefield behavior. To do this to best effect we need to situate the thinking in question and the ideological assumptions and values to connect the chain of logic to the wider context. In other words, we should look at the ways in which the wider worldview, as well as the structured environment, of the individuals and groups involved informs and bounds the way they formulate strategy and tactics.

In an attempt to do this, Katzenstein developed his idea of constitutive and regulatory norms and how these impact on strategy and behavior in military organizations. He distinguishes between norms that impact on identity formation (constitutive norms), and those that "define standards of appropriate behaviour" (regulatory norms).[7] These norms inform, enable and bound actors at the same time; guiding action by defining appropriate behavior, enabling cohesion by providing a sense of identity and unity, and bounding behavior both consciously and subconsciously by excluding other paths of action as inappropriate. These norms become increasingly powerful the more they become institutionalized, that is, become submerged in the generally accepted "common-sense" shared by a group of people or society. In this sense, a norm is at its most powerful when it has been forgotten that it exists as anything other than an accepted truth. With regards to strategy, this means that the rationality of strategic or tactical action must be judged not only by its instrumental logic and outcome, but also by the appropriateness of particular action in a given context or society. Theo Farrell uses the example of the army in newly independent Ireland to make a similar point: although it had developed as a guerrilla force, and its only hope of countering its main military threat, Britain, lay in successfully waging a guerrilla campaign against the vastly technologically and numerically superior British army, it chose to organize as a conventional state army. Despite the fact that this made poor strategic sense, it fit with the expectation and dominant norm of what a professional state army should look like.[8] Even though it lacked instrumental rationality in that it was unlikely to be strategically effective, it was appropriate in that it reinforced the identity as an independent state.

To pursue similar lines of enquiry and anchor the logic of practice beyond the immediate correlation between means and ends Bourdieu's theory of practice is a useful tool.[9] Although Bourdieu is well known and widely applied in sociology and anthropology, in strategic studies and International Relations more broadly his work is largely unknown. In effect, his theory of practice offers a theoretical framework for doing what Smith and Katzenstein suggest. It helps us to structure the relationship between influences that combine to shape practice and, in the case of religion and magic, separate out the chain of logic that underpins its incorporation in behavior during war. It also allows for recognition of appropriate behavior even when that behavior is not consciously derived through reason. In Bourdieu's words, "one can refuse to see in strategy the product of an unconscious program without making it the product of a conscious, rational calculation."[10]

Bourdieu's Theory of Practice

Bourdieu offers what he categorizes as a constructivist structuralism, or in his words, "a way of escaping from the choice between a structuralism without subject and the philosophy of the subject."[11] His theory of practice revolves around the mutually constitutive relationship of what he calls *habitus*, *field*, *capital*, and *power* to form *interests* and *strategies* and to create particular practices. Habitus is the total ideational environment of a person, "a system of durable and transposable dispositions which, integrating all past experiences, functions at every moment as a matrix of perceptions, appreciations, and actions."[12] It is the shared experience, knowledge, beliefs and outlooks of a person and in extension group, which creates the underlying context for shared meanings and practices:

> In so far—and only in so far—as *habitus* are the incorporation of the same history, or more concretely, of the same history objectified in *habitus* and structures, and also objectively concerted and endowed with an objective meaning that is at once unitary and systematic, transcending subjective intentions and conscious projects, whether individual or collective. One of the fundamental effects of the harmony between practical sense and objectified meaning (*sens*) is the production of a common-sense world, whose immediate self-evidence is accompanied by the objectivity provided by consensus on the meaning of practices and the world, in other words the harmonization of the agents' experiences and the constant reinforcement each of them receives from expression—individual or collective (in festivals, for example), improvised or programmed (commonplaces, sayings)—of similar or identical experiences.[13]

But the habitus is not a primordial intrinsic value-set, but rather a constantly evolving set of mutually constitutive relationships between members of a society, their past, and external and internal pressures. Nor does it, inclusive as it is, stand isolated. Instead it is constituted by and constitutive of the space in which it exists and the agents that share in it. Bourdieu refers to this space or context as *field*. The dispositions of agents (habitus) in a particular context (field) give rise to a set and hierarchy of resources available to the agents when acting in that context. These resources are referred to by Bourdieu as *capital*. Capital is that which is valued and provides *power* within the field.[14] In different fields, capital will therefore take different forms. Different meaning and value will be associated with practice and the "common-sense" will be interpreted differently depending on the shared frame of reference of the agents that inhabit it. But conversely, the structure created through the evolution of a field is shaped by the commonsense world objectified through habitus. It is this mutually reinforcing relationship that dictates the *logic* of practice and choices.

In an interview published in the Dutch journal *Sociologisch Tijdschrift*, Bourdieu noted that "a researcher's contribution can consist, in more than one case, in drawing attention to a problem, to something which nobody saw because it was too evident, too clear, because, as we say, it was 'staring them in the face.'"[15] Bourdieu's theory of practice is in a way evident, and could be said to be staring many analysts of behavior in the face. It represents a commonsense approach to practice generated in human societies and groups, and the apparent contradictions and paradoxes inherent in so much of it. One such paradox that is of particular concern to the enquiry pursued here is how "types of behavior can be directed towards certain ends without being consciously directed to these ends, or determined by them."[16] That is to say, behavior can have certain functions without those functions being the conscious aim and nonetheless that behavior exists in part because of these functions. This is particularly evident when considering the complex interaction of supernatural belief, ritual practice, and tactical military behavior. As will become clear in the case of the Kamajors in Sierra Leone, objectively recognizable functions such as increased combat motivation or group cohesion are derived from ritual practice informed by commonly held beliefs, even without that function necessarily being the conscious *primary* aim of the ritual practice. Similarly, the choice to deploy fighters with magical tasks such as the deflection of enemy fire is not made primarily to intimidate the enemy, although the sight of the fighters performing this task often has that effect on enemy troops. The challenge made evident by this is to account for the practice without immediately yielding to an interpretation based on elitist manipulation. In other words, to assume that if increased combat motivation was indeed one of the functions of the performance of the ritual, or intimidation a function of the deployment tactic, then those functions *had* to have been the primary aims of those who designed the rituals and practices and decided to use them at a given time. Using Bourdieu's approach we can, however, explore how the habitus

> produces strategies which, even if they are not produced by consciously aiming at explicitly formulated goals on the basis of an adequate knowledge

> of objective conditions, nor by the mechanical determination exercised by causes, turn out to be objectively adjusted to the situation. Action guided by a "feel for the game" has all the appearances of the rational action that an impartial observer, endowed with all the necessary information and capable of mastering it rationally, would deduce. And yet it is not based on reason.[17]

So at the same time as some choices are made and strategies put in place based on a rational consideration of the factors involved, not all factors will be considered with equal scrutiny. Some will be taken as given, and some may not be considered specifically at all, but nonetheless be incorporated in the practice put in place. To illustrate this we can preview the coming discussion in the case study and consider the example of the incorporation of ritual in warfare as seen also in chapter 3.

The practice of immunization rituals by the Kamajors in Sierra Leone—and indeed the Holy Spirit Movements in northern Uganda or Renamo in Mozambique—consists of the performance of a ritual and the administration of magical protection against enemy fire before going into battle. The primary aim of this practice is to *make the fighters immune to enemy fire.* Although skeptics would comment that the primary aim is to make fighters *believe* they have become immune to bullets, evidence points toward a general acceptance amongst both immunizers and those immunized of the possibility of acquiring supernatural protection against enemy fire, which in turn suggests that the aim is in fact to provide, in real terms, that protection.[18] Therefore, the practice of the immunization ritual prior to battle is the result of the relationship between the *habitus* of the Kamajors that incorporates belief in spiritual power, the *capital* of spiritual knowledge and power—necessary to carry out the ritual effectively and with the desired result—and the *field* of armed conflict. The underlying belief in spiritual power was part of the habitus of the Kamajors and in this case also of many Sierra Leoneans generally. This belief underpins knowledge of how to harness spiritual power as capital (resource), which only makes sense in so far as people believe in the existence of that power in the first place. Another significant part of the habitus is the secret society tradition, and the dispositions reflected by it, deeply entrenched in Sierra Leonean society. The secret societies constitute an organizing principle for the local communities and a structure for relaying knowledge to new generations and offering a continued network for guiding and actualizing activity in that community.[19] The main field is the armed conflict. A second field that combines with this is that of the secret societies. The two fields come together as a result of the organizing structure and power associated with the societies and the crisis brought on by the war: as rebel attacks threatened the survival of communities in southeastern Sierra Leone, hunters joined with herbalists and traditional religious leaders to combine their skills in a new society structure to counter the threat. The hunting society is itself constituted in part by the belief in spiritual power that informs its structure and organization. The disposition of the Kamajors is therefore toward drawing on the capital (spiritual power) that in the context of the habitus (which incorporates spiritual belief) has the greatest value. As a result, they choose to perform a ritual that will immunize them against bullets by bestowing spiritual protection on fighters.

The practice only makes sense with reference to the interaction of the field with the habitus and capital—in this case, spiritual belief. If there was no shared underlying belief in spiritual power, secret societies of the kind that are prevalent in Sierra Leone would not be significant.[20] But as it is, it becomes a natural point of reference for agents in the particular fields of both war and the secret society machinery for determining the choice and form of action (practice) pursued. Spiritual power and the knowledge of how to harness this power can only be understood if the role of spiritual belief is acknowledged. In this case, spiritual practice amongst the Kamajors therefore tells us something about the habitus and capital engaged, but the habitus and capital in turn tell us something about the form of behavior adopted by the Kamajors. Both directions of enquiry are useful.

In relation to the means-ends relationships that are so central to strategy and tactics, Bourdieu's theory of practice is useful in that it helps us understand the *logic of the choice of action*, or creation of practice, that underpins the means-ends process. That is, it shows why a particular means is thought to be suitable for a given end, even when the means and ends themselves go beyond the material (as in the case of spiritual beliefs). And in so doing it in a way also designates instrumental rationality—or function. But as opposed to limiting the explanation to a one-dimensional functional*ism* it allows us to explore the interplay of the agents' dispositions, the space in which they operate, and the resources that are available to them without assigning primacy to one over the other. Or to repeat Bourdieu's words, it offers "a way of escaping from the choice between a structuralism without subject and the philosophy of the subject."[21]

In warfare, we can therefore look to activity and practice directly reflecting, or influenced by, spiritual belief without making it either peripheral or central, and instead see it as a significant constituent part to behavior. This is particularly clear when examining the area where belief can perhaps have the most impact, as a motivator and reference point for existential explanation. As we have seen in the previous chapter, and will return to in subsequent sections on Sierra Leone, one of the key functions of magic and spiritual belief is mobilization and motivation. But it is by no means the only way in which spiritual belief impacts on warfare, something which will also become clear with the study of the Kamajors. In seeking to understand how and why spiritual belief is drawn upon in different forms at different times, the interplay of habitus, capital, and field is instructive. The existential framework of relevance in a given habitus becomes the natural point of reference for the agents operating within it and becomes a motivational tool because it links with that particular habitus.

Bourdieu and This Study

The Bourdieuan framework has many merits when looking at the incorporation of apparently culture-specific practices in warfare. It helps us to avoid conclusions that reduce all human activity to the bare bones of a means-ends relationship with no room for nuances. It also allows us to do justice to the choices of actors in all their human complexity while nonetheless retaining a focus on the logic and patterns inherent within them.

Mike Williams, attempting in a similar way to consider the impact of culture on security relations internationally, looked to Bourdieu to develop a framework for analysis.[22] His attempt to move beyond a constructivist–structuralist dichotomy that gives primacy to either agent or structure and to find a more inclusive way of considering choice and action led him to Bourdieu. Williams uses Bourdieu's theory of practice to evaluate the role of culture in the changing nature of security relations, and particularly the ways in which culture has gained increasing significance in the security "field" (in the Bourdieuan sense of the word) since the 1980s and shifted focus from considerations of material military hardware toward Western cultural preeminence and its universal applicability.[23] In the same way that it can aid our understanding of the influence of culture and culture-specific ideas on the approach to international security in the West, it can help us position practice and its influence in settings other than our own in relation to the means-ends relationships that determine function. These means-ends relationships are central to the analysis of warfare, but by contextualizing the "means" and highlighting the factors and processes that precipitate their choice or particular appearance we can enrich our understanding also of particular forms of tactics and strategies. For this reason it is a useful tool for understanding the ways in which something as alien (to a Western observer) as spiritual belief and practice become incorporated in wartime behavior, and why contemporary African wars are riddled with magical practice.

To apply these tools to any particular case we have to consider (1) the field and the interests that arise in it, (2) the habitus and the dispositions that are encapsulated by it, and (3) the capital that became available as a result of the habitus and field to address particular interests. In relation to the Kamajor case, which we will turn to next, this means that we have to look at (1) the context and course of the war, and the evolution of the civil defense effort, (2) the way the cosmology fit into the habitus of the population concerned—fighters and civilians, and (3) how magic became a resource—capital—because of this connection.

Part II

CHAPTER 5

NOTES ON THE CASE STUDY

Investigating something as intimate as secret society practice is a delicate matter. Doing so in the aftermath of conflict and while the society in question is being investigated for war crimes presented some challenges. Biases are common in any first-hand account of combat and especially so when the activity during the war was under close scrutiny not only by the Special Court but also by the Truth and Reconciliation Commission, researchers and local political hopefuls. As a result, many of the former Kamajors I spoke to were keen to tell their story and to "set the record straight," especially vis-à-vis the Special Court accusations. They particularly wanted to make it clear that they were fighting for the elected government of Tejan Kabbah and were a legitimate fighting force with right on their side.

Kamajor ex-combatants were generally happy to discuss their powers and successes in battle (especially thanks to these powers) but would often be at pains to explain how the more gruesome rituals described in the media or by the Special Court prosecution were exaggerations or outright lies. Other sources have, however, documented convincingly the actual occurrence of also the more brutal manifestations of ritualistic (or imitating of ritual) behavior on the battlefield, usually involving the taking of human life. But the purpose of this study was not to document war crimes or to pass moral judgment on particular activities. Setting out to investigate the way the habitus of fighters impacted on their strategies and tactics, the interrelationship between beliefs, interpretation and practice lies at the heart of the enquiry. As a result, the perceived meaning of the rituals and the belief in their occurrence, both amongst initiates and armed opponents and civilians, is as important as the fact that they took place.

The testimonies given at the Special Court proved helpful as a point of reference, especially when testimony was offered on events and details not directly relating to the prosecution's charges. In the often lengthy testimonies, witnesses would be asked to comment on specific events, relationships between members assumed to hold key positions in the CDF and aspects of everyday life in the movement. Although ritual initiation and the use of supernatural power formed a part of the prosecution case primarily against Allieu Kondewa (the "High Priest"), most of the references made by witnesses to these practices were in relation to questions about a particular battle or the personal involvement of the witness. Some of the testimonies were observed by me in person at the Court, others seen on video or read in trial transcripts.

Finally, insights on a group's behavior can also be found through the accounts of non-members that interacted with them. To this end, civilians and military personnel involved with the Kamajors were also consulted. Particularly interesting were the accounts of the former Executive Outcomes employees who fought alongside the Kamajors at different stages during the war.

Fortunately there is a growing and insightful literature on the war in Sierra Leone that both helped give the history of this "messy" war and perspectives on the actors and their behavior that complemented the original research undertaken for this study in crucial ways. Particularly invaluable were the works of Danny Hoffman, who spent a significant amount of time with the Kamajors in the later years of the war, and David Keen, who has painstakingly documented the course of the war and its actors. The documentation created by the Truth and Reconciliation Commission and NGOs like No Peace Without Justice also captured important details and testimonies that helped underpin this study.

Relying on these sources for inputs this study is, however, not another history of the Sierra Leone war or an exhaustive consideration of the use of ritual and magic by the actors in the war. It is a specific consideration of one group, the Kamajor component of the Civil Defense Forces, to put their behavior on the battle field in context. Although a crucial part of this is their interaction with other key players in the war the main focus remains with the Kamajors.

CHAPTER 6

THE SIERRA LEONE CIVIL WAR AND CIVIL DEFENSE

At the start of the war in 1991, the Sierra Leone Army (SLA) started to make use of local hunters as scouts during army patrols, drawing advantage from their knowledge of the local terrain. Simultaneously, local communities began to organize defense groups to fill the gap left by the underfunded and poorly equipped army that was providing poor protection from the advancing rebel forces. Like any war, the civil war in Sierra Leone was the result of a series of interconnected internal and external developments, and the specific form it took reflected these. To understand why the Kamajors became key players in the war and why chose to incorporate magic and traditional beliefs in the conduct of their civil defense, we must understand both how the war played out and how civil defense was organized. These developments form an important part of the field and over time came to impact also on a habitus shaped in part by war.[1]

THE ORIGINS OF WAR

Sierra Leone's descent into civil war stemmed from a combination of decades of misrule that had eroded state institutions and service, and an increased alienation of the non-urban population outside the capital Freetown. Patronage systems that had been cemented during the rule of Siaka Stevens (1968–1985) could not, and would not, be got rid of by his politically much weaker successor Joseph Momoh. Instead, Momoh ended up trading economic stability and the government's overall control of the country's resources for the ability to hold on to power, allowing the clientelism and illicit trading networks to remain. In 1991, the government was challenged militarily by a segment of the alienated population that had fomented rebellion in the hinterland and in exile. This challenge would persist for more than ten years.

Analyses of the origins of the civil war in Sierra Leone have examined at length the combinations of corruption and alienation that combined to trigger the onset of war.[2] At the core lies a privatization of state resources by a small elite, in the case of Sierra Leone particularly the diamond fields in the east and access to the regional and international markets that came with them. Through this central income-generating resource, and its control by Stevens and in part Momoh, patronage was distributed and loyalty secured. Those willing to cooperate with the ruling All People's Congress (APC)—in effect supporting the system as it had been created—were rewarded, and those voicing dissent were excluded or punished.

But in addition to creating grievances through exclusive patron-client relationships, the informalization of the economy also led to direct economic grievances as the government's capacity to raise revenue from state enterprises was reduced and state services were eroded to the core.[3] At the end of the cold war, the clientelist structures were shifted from local clients to embrace foreign firms, removing the financial rewards even further from the Sierra Leonean setting and breaking the patronage systems that in spite of their exclusive nature had reached further down into society.[4] As a result, unemployment rocketed and the options available especially to the younger population grew increasingly limited. Without an accessible state structure, or even private sector, they were denied any form of social mobility.

Dissenting youth was therefore readily available to the Revolutionary United Front (RUF) and its patron Charles Taylor in neighboring Liberia. The former hoped to rectify the perceived injustices of the patrimonial system that denied them access to power and material wealth—part of this agenda incorporated private ones of revenge, notably a desire by RUF leader Foday Sankoh to punish the APC regime that has jailed him for seven years for his alleged participation in a coup plot in 1971.[5] The latter, saw an opportunity to secure income for his rebel National Patriotic Front of Liberia (NPFL) through resource extraction in Sierra Leone, as well as to challenge the Sierra Leonean government over its support for and decision to host ECOMOG peacekeeping forces bound for Liberia.[6] Supporting the RUF allowed Taylor to do both. For the RUF rebels, war offered a chance to generate income through looting and subsequently control of diamond fields, but also the prospect of replacing the ruling elite in Freetown and turning the system that had so far deprived them of opportunities into their own.

Over a decade of war the motivations of the actors that were or became involved evolved. Parts of the army began to collude with the rebels and found an interest in keeping hostilities going to allow them to benefit from lucrative diamond extraction as did the rebels. Civil defense militias, as we shall see, became increasingly involved as the war progressed—primarily for community defense but parts of the militias also found an interest in enrichment and increased personal power that came through war. Liberian groups came and went on the Sierra Leone side of the border, replicating some of the animosities from the Liberian civil war in the Sierra Leonean conflict dynamic; supporters of Taylor (notably his NPFL) would have dealings with the RUF and those opposing him (such as ULIMO) sought support from the army and civil militias. Finally external actors, including ECOMOG and private military companies were drawn in to the conflict and its logic of power and the allure of wealth.

THE COURSE OF WAR

When the RUF incursion from Liberia began in March 1991, the 300 rebel fighters (made up in part of Liberian NPFL fighters and others from Burkina Faso) gained ground quickly from the SLA whose 3,000 troops were ill-equipped and poorly prepared.[7] In spite of RUF leader Foday Sankoh's inability to secure any large-scale support from the local population—largely due to the force's brutal methods—by July 1991 they were an entrenched presence in the country and held their ground.

President Momoh's inability to defeat the rebels inspired an army coup that overthrew him in April 1992. The government was replaced with a military junta going by the name the National Provisional Ruling Council (NPRC) under the leadership of Valentin Strasser, a former ECOMOG soldier in Liberia. The NPRC were allegedly supported by the RUF in its bid to oust the APC government on the understanding that they would be invited into the new government.[8] However, although the NPRC made some peace overtures to the RUF they also stepped up efforts to defeat them militarily. The rebels in turn also quickly abandoned the idea of peace. They were under pressure from Taylor and the NPFL in Liberia who needed the continued income generated by the RUF's control of diamond fields in Sierra Leone, and following the Liberian lead and the reinvigorated NPRC offensive they opted for a continued war effort. Quickly the Sierra Leonean soldiers also became involved with looting to supplement their meager (and often unpaid) salaries and regular abuse of civilians.[9] So the war continued. But despite the collusion of some of the armed forces, others made a significant effort against the rebels and the army under the NPRC went on to make significant advances in the year following their accession to power.

In 1994 the RUF changed its tactics to counter the success of the NPRC and spread from their predominantly southeastern bases across the country, involving almost all of Sierra Leone in the war. After a year of increasing losses for the NPRC, including the forced closure of key revenue earning mines at Sieromco and Sierra Rutile, in 1995, the RUF reached the outskirts of Freetown signaling the government's inability to avert the threat decisively. The NPRC turned to the outside world for help. Initially Strasser invited British security company Gurkha Security Services to help his army quell the rebels. Despite some initial successes the Gurkhas withdrew after their commander was killed in battle. To replace them, Strasser extended an invitation to Executive Outcomes (EO), another private military contractor, to boost the state's fighting forces. EO provided support to the army and to the local civil defense militias, mobilized in response to the increasing insecurity brought about through the RUF insurgency. The Kamajors were one of these regional civil defense militias that was mobilized to counter the rebel onslaught. The combined effort of EO and their civil defense militia allies saw the recapture of the main diamond and mining fields and seriously challenged the RUF.

But the regained advantage did not hold as political quarrelling took priority and allowed the war to gain new wind. Argument over Strasser's political future and national elections led to another coup in which Strasser was replaced by the deputy head of the NPRC, Brigadier-General Julius Maada Bio. Bio, despite initial opposition to elections was forced to cede and called elections that brought Ahmed Tejan Kabbah to power in early 1996. Once in power, Kabbah formalized the role of the civil defense militias that had been employed by the NPRC. Following the elections, a series of negotiations took place attempting to reach a political settlement with the RUF and return peace to the country.[10] The Abidjan peace agreement was signed in November 1996. Part of the agreement was a promise to end the contract with EO and effectively take the RUF's main military challenger at the time out of the picture. But despite the signing of the agreement peace remained elusive. In May 1997, Kabbah was overthrown by Johnny Paul Koroma, a Major in the SLA and a military junta, the Armed Forces Revolutionary Council (AFRC), again took over government. The soldiers that made up the junta repeatedly quoted the humiliation of seeing the president place his trust in civil militias rather than his armed forces as a key motivation behind the coup and many of the soldiers that had been colluding with rebels or engaging in their own illicit activities were no doubt seeing these activities curtailed by the increasing civil defense efforts. One in power, the AFRC immediately extended an offer of truce and cooperation to the RUF and began a year of rebel-junta collaboration.[11]

Unrecognized internationally, the AFRC was challenged by sanctions imposed first by ECOWAS and later the UN, and the subsequent remandating of the Nigerian peacekeepers already in the country as ECOMOG II.

Forcing the hand of the Junta in this way, another agreement was reached that called for the reinstatement of Kabbah as president.[12] In the end, following intensified fighting between the AFRC/RUF and the Civil Defence Forces (CDF) in the provinces and continuing clashes between AFRC/RUF and ECOMOG troops, Kabbah was restored to power by the Nigerian peacekeepers and their civil defense allies in March 1998.

But again the war did not end. Instead, fighting returned to Freetown on an unprecedented scale in January 1999. The AFRC/RUF invaded the capital and for two weeks fierce street fighting engulfed the city and more than 5,000 people, primarily civilians, were killed and many more severely mutilated. The last stage of negotiations followed with yet another attempt to bring the AFRC/RUF into the government and out of the bush. A new peace agreement was signed in Lomé, UN peacekeepers were brought in, and AFRC/RUF leaders were given senior government positions.[13] The new order lasted barely a year. In May 2000, following another eruption of violence in Freetown, RUF fighters attacked UN troops in the countryside and abducted several hundred UN peacekeepers, prompting a British military intervention that in addition to securing the evacuation of foreigners from the capital severely damaged the rebels.[14] Following the intervention, Britain committed to staying in the country to train the armed forces.[15] Also in 2000, neighboring Guinea took a more proactive approach to the war as it claimed Charles Taylor was directing RUF attacks into Guinea from across the Sierra Leone border. In response they bombarded Sierra Leonean border towns thought to be RUF strongholds and inflicted heavy casualties on the rebels.[16] Following the embarrassment of having UN peacekeeper repeatedly taken hostage, the UN strengthened the peacekeeping force and its mandate, increasing its numbers from 11,000 to 13,000 at the end of May 2000 and again to 17,500 in March the following year.

The greater and more proactive international military presence in the country—UN, British and Guineans—helped to restore calm and the process of establishing peace could slowly begin. In 2002, Kabbah addressed the nation and announced "de war done done," marking the official return of peace to Sierra Leone.[17] To date, disarmament has officially ended, elections have been held twice, a Truth and Reconciliation Commission has completed its work, and a Special Court has tried members of the RUF, AFRC, and CDF for war crimes. A trial of Charles Taylor for his involvement in the war is under way in The Hague.

The Emergence of Civil Defense

Communities across Sierra Leone had mobilized local men for community defense in a variety of guises since the early days of the war, but some gained more prominence than others. Although the exact chronology and geography of the use of local hunters by the SLA is somewhat obscure, the Tamaboros in northern Sierra Leone (mainly ethnically Kuranko) are thought to be the original hunter militia. The first hunter militia organized specifically for the purpose of countering the rebel threat was led by Dr Alpha Lavalie, a university lecturer from Fourah Bay College in Freetown.[18] The Kamajors as we have come to know them emerged around 1995, although at that stage both local men and traditional hunters had already organized in defense units in the Mende south and southeast of the country, many of them going by the name of "kamajor," which means hunter in Mende.[19] The Kamajor initiation society was established by a group of traditional herbalists and Qu'ranic scholars in Bonthe district. They came together to help boost the power of local defense through their traditional crafts, and drew heavily on traditional and spiritual beliefs and practices. However, even before then some civil defence fighters received magical protection by local initiators. The specific nature of initiation and the Kamajor *society* will be explored in greater detail in chapter 7.

There were four civil defense groups in addition to the Kamajors; the *Tamaboro*, *Donsos*, *Kapra*, *Gbethis,* and the Organized Body of Hunting Societies (OBHS) in the Western area. The different names are simply the different words used for hunter in the language spoken by the particular group. In the Western Area, including Freetown, the English name makes this clear. Similarly, in the north "tamaboro" can be translated as meaning "walk-about bag" in Kuranko, signifying the hunter practice of saying they "go walkabout" rather than hunting.[20] All except the Gbethis had existed in some form before the war and were structured around hunting and, where they existed, hunting society traditions. In many cases—as will become clear as we look more closely at the Kamajors—the hunting tradition was more of a rallying point than a solid organization easily converted to a wartime militia. The groups were defined largely by ethnicity and were geographically distinct, although the extent to which they existed as distinctive and pervasive identities before the war is debated. The Kamajors are Mende from the south and east; the Tamaboros are made up of Kurankos, Yalunkas and Susus from the north and northeast; the Donsos are Konos from the East; the Kapras are Temnes from the north and northwest, and the OBHS was made up of a number of groups resident in the Western Area including Freetown, Waterloo and Lumpa.

The Kamajors became the most widely known and were also the largest of the civil defense groups in the country. Their numbers are disputed but disarmament figures put the total around 37,000 by the end of the war. It is, however, very hard to accurately assess the number of combatants involved with the different groups. This is in part because membership shifted as fighters defected from one group to join another and in part because membership of a group could mean a noncombatant role, such as cook or porter. Figures from the demobilization process give an indication of the rough numbers involved, but these figures are compromised by the fact that not all combatants went through the demobilization process, some went through it twice and some went through it without having been a combatant. Nonetheless, looking at the UN figures for demobilization at the end of the war, it is clear that the CDF together with its nemesis, the RUF, occupied the central positions in terms of sheer manpower, even if large parts of their numbers no doubt included noncombatant members (see Table 6.1).[21]

The transformation of the local defense units that sprung up in the early days of the war to the nationally coordinated CDF proceeded through a series of more or less distinct phases. Although these phases overlap, they can be said to correspond roughly with the time periods 1991–1994, 1995–1996, 1996–1997, and 1997–1998.

The first phase (1991–1994) saw the emergence of local civil defense units. As the army proved unable to defeat the RUF local resistance groups began to form. Patrick K. Muana describes how precedents were set as early as 1991–1992 when local vigilantes were organized into fighting forces by Captain Prince Ben-Hirsch in Segbwema, and by Dr. Alpha Lavalie and the Tamaboro.[22] Lavalie's group drew specifically on hunter traditions and was aided by his status as the political leader of the main opposition Sierra Leone People's Party (SLPP), which held the support of many of the people who had been displaced by the war. But both relied on a good relationship with the traditional leaders that provided the wide support necessary for recruitment.[23] Operating check-points and organizing look-outs, the local defense groups provided a sense of security to local communities that the SLA was incapable of offering, in addition to presenting a first line of defense against rebel attacks. The defense units were equipped with mostly cutlasses and single-barrelled shotguns, making up for what they lacked in weaponry with a fierce will to resist the rebels and protect their communities.[24]

Table 6.1 Disarmament figures for the RUF, CDF, and AFRC/ex-SLA (data from UNDDR Resource Center)

Armed group	Number participating in disarmament
RUF	24,352
CDF	37,377
AFRC/ex-SLA	8,527

In the Kamajor heartland—the primarily Mende South and East of the country—men began to organize for local defense around the same time. However, in the early days of the war many of the Mende men who would later become Kamajors had joined the rebels and would only return and change their allegiances as the aims of the RUF became increasingly indistinguishable from mere enrichment and their methods almost exclusively focused on the brutal targeting of civilians. Notably, in 1994 almost an entire group of RUF rebels known as the "Action Group" abandoned the RUF and later rejoined the war on the side of the civil defense.[25] Having originally joined the rebels to challenge government corruption and the APC iron-grip on the political system, this same desire to protect their livelihoods and communities from exploitation and abuse led them to abandon the RUF for civil defense. As the "sobel" phenomenon of either direct collaboration between SLA soldiers and RUF rebels or the indulgence in rebel like activities by SLA soldiers grew, the distrust in government protection also increased and the idea of civil defense became increasingly popular.

The second phase (1995–1996) saw the civil defense cooperation with the SLA intensify; initially through the combined efforts of civil militias and EO at the behest of the NPRC government and after the election of Kabbah through the official government sanction of a national civil defense force. During this time the Kamajor society began its "formal" organization as a civil defense militia. Formal is, however a word that should be used with caution lest it suggest the formation of a highly centralized and hierarchical military organization. What happened in 1995 was that the support and training the civil defense militias received through their cooperation with EO, combined with a more proactive use of initiation of new recruits allowed the fighting groups to take on a more unified identity and become more effective as a fighting force. Although civil defense in the South and East already existed, the Kamajors emerged in their characteristic guise in 1995 when organized ritual initiation of recruits began in Bonthe District.[26] Previously, Kamajor fighters would have received some ritual protection by individual initiators in their locality and would have been organized by local leaders. Chief Samuel Hinga Norman, who later became a head of the national civil defense organization, presided over one of the most effective civil defense militias in his chiefdom Jiama Bongor before he took a more proactive role in the larger civil defense effort. Following the expanded role of these fighters from 1995, the initiates proceeded to recruit and train additional fighters in the South and East, and rapidly swelled their ranks. A system of chiefdom level recruitment and training further assisted their mobilization and helped anchor the militia with the communities in the region. By 1996 almost all chiefdoms in the south and east of the country had Kamajor societies and recruitment mechanisms, and national coordination of the civil defense units was formalized through the appointment by President Kabbah of Hinga Norman (also made deputy defense minister in the newly elected government) as National Coordinator of the Civil Defence Forces.[27]

Although the CDF remained largely a label at this stage, they became more closely associated with the Kabbah government, particularly in the eyes of the army that perceived them as a threat to their own position as the national security force. The transformation of the civil defense units into a more organized structure was significantly aided by the arrival of private military contractors EO in 1995. EO ended up using primarily Kamajor fighters in their operations as they proved both more motivated, reliable and militarily capable than their SLA counterparts.[28] The training and organization that came with the EO collaboration helped solidify the militia and cement its role also after the departure of EO.[29]

During the third phase (1996–1997) the Kamajors and the CDF grew stronger but at the same time the SLA began to cooperate with the rebels on a greater scale than before. The discord between the CDF—primarily its Kamajor component—and the SLA increased and culminated in the complete break of cooperation when the AFRC took power. By 1996 the Kamajors had become an effective fighting force and a clearly defined party to the conflict, but their preferential treatment by the government sparked discontent in the ranks of the armed forces and the Kamajors were in part blamed by the AFRC for the coup in 1997. AFRC leader Johnny Paul Koroma specifically named the Kamajors in his justification for challenging the elected government:

> the SLPP tribal hunter militia, the *Kamajors*, received logistics and supplies far beyond their immediate needs. This was enough indication of the preference for the private army over our Armed Forces, foreshadowing the ultimate replacement of the Constitutional Defence Force by Mr Kabbah's hunters.[30]

In addition to receiving support from the Kabbah government that appeared disproportionate in view of the poorly equipped SLA, the Kamajors were also associated with the governing Sierra Leone People's Party because of the party's traditionally Mende power base in the south and east of the country. Government support for the Kamajors was therefore seen not just as a military strategy drawing on local defense potential but an ethnically driven policy of building an armed force loyal to the SLPP leadership. As most of the army, and particularly its higher ranks, was northern in origin, the threat of an ethnification of state security that favored the south was a personal threat to many army officers. The AFRC's open cooperation with the RUF as well as its overthrow of the elected Kabbah government not only cemented its antithetical position vis-à-vis the Kamajors but set the scene for the intense fighting that followed during the year of AFRC rule.

Phase four (1997–1998) saw the Kamajors and the CDF more broadly emerge as the main adversary for the AFRC/RUF, fighting alongside ECOMOG to reinstate the Kabbah government. National coordination of the civil defense effort had been given impetus in April 1997 when parliament agreed to legitimize the use of arms by hunter civil defense units, but the process of central organization that was sanctioned in this way was halted by the coup that followed in May. It was from exile in Conakry that President Kabbah and his government institutionalized national civil defense in the form of the "Civil Defence Forces," under the coordination of then Deputy Defense Minister Hinga Norman. Norman was given the position because of his already close links with the Kamajors and because of his military background, which it was thought would allow him to interact more effectively with the fighting forces.[31] At this stage, national coordination of the CDF in one sense became further centralized as a main base was established and a War Council set up to streamline operations.[32] However, in many ways this was more of a façade than a reflection of what was actually going on in the civil defense militias around the country. Operations were still fragmented and tactical decisions were practically all devolved to local fighting units. The civil defense groups did, however, actively seek to boost their ranks to face the AFRC/RUF challenge and extended their recruitment beyond Sierra Leone and drew on both Liberian mercenaries and Sierra Leoneans residing in refugee camps in Guinea.[33] Some Liberians or Sierra Leoneans with mixed Liberian parentage were already fighting alongside the militias but during the AFRC year in power they were joined by many more fighters with experience from the Liberian civil war. They also received explicit assistance from the Liberian ULIMO group. Because of their experience in other wars, Liberian fighters tended to have more extensive weapons training and helped form special more professionalized units within the CDF.[34] Most of these fighters were based on the Liberian border at Base One—the original CDF base at Gendema-Bo Waterside where Eddie Massallay called the civil defense militia fighters to gather following the coup.

In December 1997 the CDF launched its most concerted campaign against the AFRC/RUF, attempting to prevent transportation in and out of Freetown as well as severing communications links between the enemy forces so they could be engaged in isolated battles. The operation specifically targeted main roads, seeking to control the transport routes and to fragment the junta's military capacity. This operation, known as Black December, saw widespread destruction and brutal methods on the part of the CDF. In March 1998 ECOMOG and the CDF together drove the AFRC/RUF out of Freetown and reinstated Kabbah as president. After the AFRC had been driven out of office the CDF continued to fight alongside ECOMOG on the side of the government against the AFRC/RUF.

THE DIFFERENCE BETWEEN THE KAMAJORS AND THE CDF

Although the Kamajor society made up the largest body of fighters in the CDF, the two cannot be entirely equated. Nor can their names necessarily be used interchangeably. The main difference between the two was in nature—the Kamajors were a secret society, and the CDF an official governmentally sanctioned fighting force. Although both comprised of fighters, many of whom were the same, the Kamajor society remained within the tradition of secret societies as it existed in the Mende south and east of the country, and largely drew its strength from that tradition (we shall return to the specifics of this tradition in chapter 6).

Members of the Kamajor society were bonded together through their initiation experience and their sense of loyalty and ethics stemmed from this bond (as will be explored in chapter 7). The structure of the society and dealings within the society was for members only and did not lend itself to straightforward cooperation with an external party let alone a government expected to operate with a degree of transparency and accountability.[35] To facilitate the collaboration between the government and the Kamajors (and indeed the other hunter militias that also drew on secret society traditions) the CDF was set up as an umbrella organization. Many but by no means all of the CDF leadership were initiates of the Kamajor society. For example, Vice-President Joe Demby was involved both with the initial civil defense efforts in Kenema and with the Kamajor movement but was, according to his own testimony, not initiated.[36] But in their role as CDF they could liaise with both government and parliament in an official capacity, while continuing to represent unofficially the secret society (and without breaking any of the society rules by speaking of its activities to non-initiates).

Many Kamajors saw the CDF as disruptive to the success of the Kamajors and their disciplinary structures in particular, claiming that the national institution did not instill the same sense of loyalty and obedience as the society did and as a result compromised the successful force already built.[37] We will return to this dilemma below when considering the role of initiation in recruitment and, crucially, for discipline.

Although informal coordination of civil defense had begun in 1996, the CDF only became an official entity in 1997, and effectively a national organization only after the overthrow of the AFRC in 1998. For this reason it only makes sense to talk of the "CDF" from 1997 onward. "Civil defence" did, as we have seen, exist prior to 1997, as did the Kamajors, but it is important to bear in mind the distinction between these entities both when considering their organizational structure and the role played by magic and ritual. In effect, the "society" allowed for the creation of a motivated and respected defense capability, drawing on ritual and supernatural power. The "defense force," allowed it to fit into a broader political reality with institutions and workings that could not connect with the traditional society structure but were suitable for the open political sphere.

Official and Actual Structures of Civil Defense

The organizational structure of the CDF is a contentious issue, not least because of the war crimes trials that sought to establish who bore the greatest responsibility for the atrocities committed by the parties to the conflict. There has been a tendency to try and fit the civil defense apparatus into a neat hierarchy with a clear chain of command, but the reality is less clear-cut. This was partly because the nature of the war required a localized response and partly because of the nature of the organization itself. The CDF as a "national" organization emerged as a result of the 1997 coup and a desire by the government in exile to support a concerted effort to oust the junta. The national structure of the CDF was therefore in part a natural formalization of an existing order, and in part an external response pushed by the exiled government in Conakry. But despite overlapping layers of authority and a confusing use of titles and ranks, the civil defense effort did manage an element of coordination. However, when looking closer at the organization of civil defense and its overlapping national and local structures, it becomes clear that a central command was not the focal point of the force, and that there were links and loyalties formed by other relationships, notably the initiation of fighters into the Kamajor *society* (or indeed the other regional societies that functioned along similar lines) that took precedence.

With the rapid expansion of the Kamajor civil militia from 1995 onward their swelling numbers and systematic approach to recruitment, simultaneously required increasing centralization and localization of their effort. It required centralization in that initiators needed to be gathered and deployed to local centers where initiation could take place and localization in that this created multiple points of civil defense concentration around the country. However, after the May 1997 coup the Kamajors and other civil defense units were forced to abandon towns and go into hiding after the new junta turned the army against them. The main Kamajor—and subsequently CDF—camp was set up at Talia Yawbeko in Bonthe district, and became known as Base Zero (see map 1).

Base Zero became famous as both the Kamajor and CDF headquarters because the national coordinator, Hinga Norman (himself an initiated

Kamajor), regularly operated out of the base. But Base Zero was created in direct response to the coup in the second half of 1997, and remained the center of operations only while the AFRC was in power. Another important base had been set up earlier at Gendema/Bo Waterside on the border with Liberia, and was referred to as Base One.[38] This had been the frontline of much of the fighting in part because of its proximity to the diamond areas in the east, and in part because the border was where both sides were receiving supplies from Liberian allies. Base One also acted as an assembly point for Liberian fighters of the ULIMO faction who joined forces with the CDF against the RUF.[39] It continued to be a major second base of operations, particularly for the Liberian CDF fighters who worked closely with Maxwell Khobe's ECOMOG troops.[40]

When the AFRC seized power and declared the end of the CDF, many Kamajors retreated from the main towns and gathered to regroup from less exposed positions.[41] Many went back to their villages, but some fled to Base One or as far as Liberia from where they reorganized and decided to set up a main base of operations from which to launch an attack on the AFRC. Talia Yawbeko was chosen because of its relatively protected location—away from AFRC strongholds (generally wherever there had been army barracks) and with only one access road, yet close enough to the highways connecting Freetown, Bo and Kenema to make attack viable (see map 1). In addition to these geographic considerations, it was chosen because of the initiators' close connections with the area and the local chief.[42]

The organization at Base Zero represented the first attempt to combine the national framework and the Kamajor society in one management structure, and highlighted the inherent tension between these two layers of the movement. In the end, this led to the organizational structure being more of a theory than practice.

The new base replicated the structure that had previously been found at local recruitment centers: the initiators provided magical protection and motivational guidance, and those with military experience provided the military training necessary for recruits.[43] Allieu Kondewa became the main initiator and High Priest of the movement at Base Zero.[44] One account of why he was made High Priest, as opposed to one of the other initiators active prior to this time, explained how he outperformed the other contending initiators at a meeting in Bonthe district in 1995 and as a result became the de facto "head initiator." However, other accounts claim the initiator's circle remained flat hierarchically, and that the office of High Priest was tied to CDF organizational needs rather than the Kamajor society structure; suggesting that Kondewa's appointment was less anchored in authority than in practicality.[45] Under his official leadership, five other main initiators operated in the regions, each with additional staff and initiators. The five senior Kamajor initiators were Kamoh Dr Lahai Bangura, Kamoh Brima Bangura, Mama Munda, Kamoh Alie Sesay, and Kamoh Dr Mohamed Mansaray.[46] Although it appears the initiators all acted with a high degree of independence in their areas of operation, and as a result initiations would differ depending on the place and initiator. In addition to the initiation cadre, regional commanders coordinated and organized the CDF outside the camps at Talia and Gendema.

It was, however, at Base Zero that the CDF national hierarchy was first established. The triumvirate leadership of Hinga Norman, Allieu Kondewa and Moinina Fofana was introduced to the movement there. Although the three were largely known in their own capacities before this date, it was in the CDF hierarchy that Kondewa and Fofana were awarded the offices of "High Priest" and "Director of War." Norman already held the office of national coordinator of the CDF, and was in addition deputy minister of defense in Kabbah's government, then in exile.[47]

Although these titles hint at an organizational structure, the reality of the command chain and logistical functioning of the CDF is more obscure. Titles and responsibilities were certainly conferred, but were not necessarily indicative of actual powers. Several attempts to sketch the organizational hierarchy of the CDF betray this partial confusion. In the diagram produced by *No Peace Without Justice* (figure 6.1), the office of the High Priest does not feature at all, although it would likely have been subsumed under the label of Director of Personnel, as his role was connected to the recruitment process. It also separates the office of the Director of War from that of the Director of Logistics, although these both appear to have been headed by the Director of War Moinina Fofana. The role of the National Public Relations Officer is also unclear as Norman generally dealt directly with Fofana, Kondewa and indeed many of the commanders. However, the examination of the command and control chain in the CDF by Richard Iron—a military expert witness called by the prosecution at the Special Court to give his view on CDF organization—highlights the differential nature of some of the positions, as well as the flow of decision making and the subsequent execution of orders. Irons' depiction of the CDF's military command structure seems to represent more clearly the significance of individuals and the way the hierarchy and processes within it became shaped by the roles of these individuals. Figure 6.2 is extrapolated from Irons' explanation of the CDF military structure.

However, Danny Hoffman, in his expert witness report to the Special Court, criticizes Richard Irons' view of the organization and suggests a much looser and locally anchored movement.[48] In his view, most decision making remained local, although groups and units would approach both Base Zero and Base One for logistical support when necessary. Different

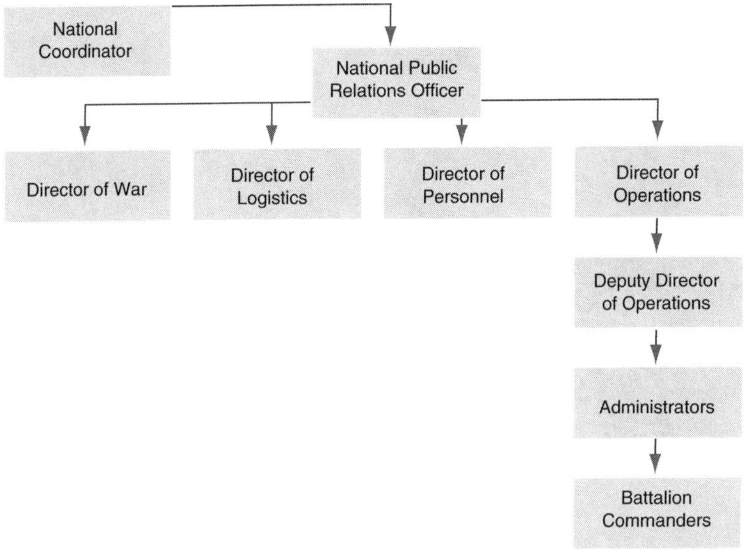

Figure 6.1 Command Hierarchy of the Civil Defence Forces in Sierra Leone, June 2007, from the No Peace Without Justice publication "Conflict Mapping in Sierra Leone: Violations of International Humanitarian Law from 1991–2002." By L. Alison A. Smith, C. Gambette, and T. Longley, 2004, available from www.npwj.org

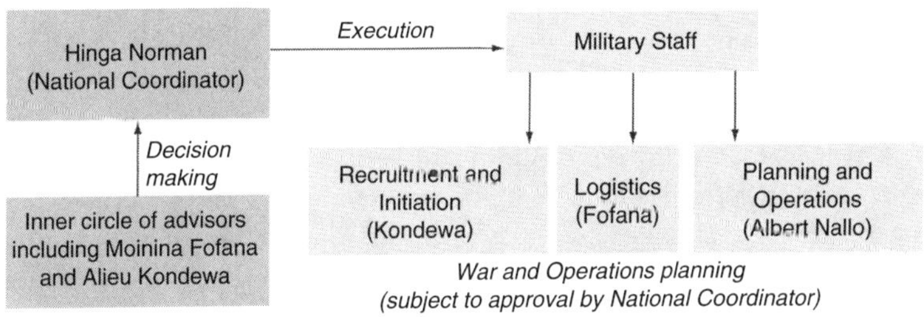

Figure 6.2 CDF military command structure (based on narrative in Iron, R. Military Expert Witness Report: 2005)

groups also maintained different types of relations with the senior leadership depending on their patron-client relations; individuals able to provide military, logistical or administrative capacity would command greater authority and secure a greater number of clients. As a result, many of them would be given (or some would take) honorific titles such as commander or general. Although this did signify a level of authority, particularly in the locality

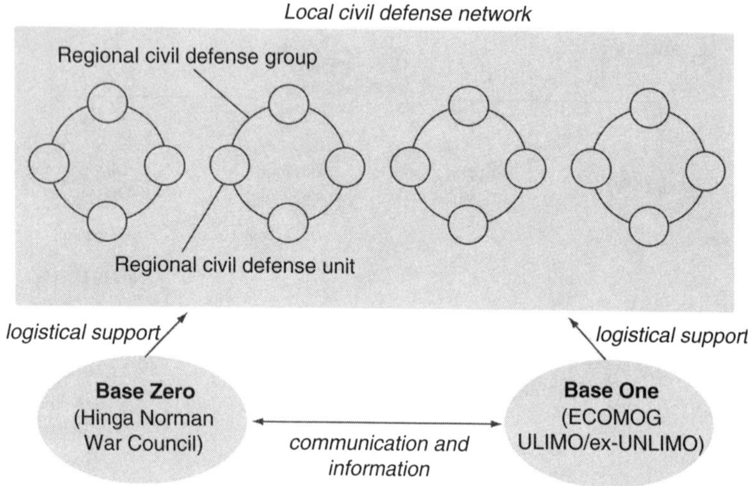

Figure 6.3 CDF organization (based on narrative by Hoffman, D. Expert Report on the Kamajors of Sierra Leone: 2006)

where the support of clients would translate into capability, it did not necessarily correspond to a predetermined military organizational structure with a rigid chain of command where duties and jurisdiction were communicated through rank.[49] Figure 6.3 is extrapolated from Hoffman's idea of a more network-based system of defense groups.

The process through which decisions became executed orders within the CDF appears to have been dependent on the type of decision and the time at which it was being taken. Additional bodies such as the War Council were set up to provide checks both on the command chain and the behavior of fighters. But on a day-to-day basis, the strategic planning and operations appear to have been less structured than the official organizational charts would suggest. The actual command chains and interpersonal relationships in the CDF were closely connected to those established through the Kamajor society—especially the loyalty and allegiance that was created between initiator and initiates. A closer look at the powerful sounding War Council reveals this in part false sense of command, as does an examination of the administration of logistics and training.

The CDF Decision-Making Apparatus and Command Flexibility

In November 1997 the War Council was established at Base Zero. The Council was based at Base Zero together with the CDF leadership until Bo and Kenema were captured by the CDF in February 1998, when it relocated to Bo. The Council was charged with making strategic decisions, such as choosing targets for attack and directing troop movements, as well as relaying information to Hinga Norman who was responsible for communicating with President Kabbah, at this time in exile in Guinea. However, the actual role and degree of authority of the Council is debated. Some have argued that ultimate power remained with Norman who could bypass and overrule the Council if he chose to.[50] Others have suggested the Council was set up to provide some greatly needed legitimacy to the movement after the large recruitment drive in 1996–1997 had compromised some of the movement's reputation through a lack of discipline.[51] And yet others have claimed that membership of the War Council, like most bodies and titles conferred within the CDF, were honorifics rather than signifiers of actual official status and level of control, and part of a network of patron-client relationships built up between those who could provide and those who would offer their services and loyalty in return for such patronage.[52]

It seems unlikely that the Council was set up merely as a cosmetic solution to poor press. More likely is that it reflected a genuine need for coordination and centralization to successfully wage the war against the AFRC/RUF. However, the type of coordination was not necessarily directly linked to military control. The trust of the local communities needed to be regained where diminished, and the way to do this successfully was through the chiefs. The chiefs were the traditional bastions of power in all of Sierra Leone, and chiefdom politics were of an overriding importance. In their jurisdictions, the chiefs were responsible for allocating resources in the community and mediating any locally arisen conflict. In terms of civil defense, the chiefs had been central in lending support to the formation of defense groups and early recruitment efforts. However, the relocation of the CDF away from urban centers after the coup, and the massive increase in initiation that had occurred even before the regime change, had created new patron-client links between fighters and commanders and to varying extents marginalized many chiefs. To enable smooth operations, particularly in the south, regional support from the chiefs became increasingly important when the CDF set out to counter the junta. By inviting some chiefs onto the War Council and relaying more information about the progress of the war against the Junta to others, closer links could be reestablished.

At the same time, actual decision making was far from centralized and continued to reside primarily with regional commands and local units. Although the official national leadership technically had the last word because of their control of weapons supplies and monopoly on national and international communication, in reality this was rarely practically possible: Even the potentially sanctioning power of the national leadership arising from their control of arms was limited by the general practice of accumulating weaponry from the battlefield and vanquished enemies, and indeed other sources of supplies such as ECOMOG, especially around Base One. In some ways, rather than central points of command, both Base Zero and Base One were what Hoffman has referred to as "locations to which one could travel or send delegates in order to ask for supplies."[53]

Despite this lack of effective and immediate powers of decision making, the Council would offer advice and anchor decisions with their constituencies. Because of the geographic location of Base Zero, this was particularly important in the south. The CDF also stood to gain increased international support with a clearer organizational structure in place, even if its function was not always true to its apparent form. The welcoming of international journalists at Base Zero suggests as much. In relation to the schematics of the CDF hierarchy given above, the War Council would therefore have floated alongside the group consulted for decision making, adding an additional source of input, although one with no executive power.

In theory, once strategic decisions had been made by the War Council, they would be relayed to Norman who would liaise with the Director of War, decide whether to follow the recommendation and issue orders to the relevant commanders.[54] Although the Director of War was technically in charge of tactical deployment, reports have suggested that the High Priest retained a high degree of control over the commanders he had initiated and often directed them personally. Other testimony suggests that both strategic and tactical specifics were generally left to individual commanders. One Kamajor testifying to the Special Court commented on this porous chain of command that

> There was no one individual that was in control of the deployment, because I was thinking that could have been the role of the Director of War. But the high priest was doing his deployment. Every commander would just say, "Well, can we have—to be deployed in that area." And, you know, something like that. So almost anyone who could command up to ten or 15 men will just think of an area, say, "Okay, we are going to deploy there."[55]

Similar accounts have been given by other commanders who claim they were given the freedom to conduct their missions (generally only broadly defined by the central command) as they judged best.[56] This practice was the result both of the constraints on communication between Base Zero and the commanders spread across the southeast, and of the acknowledgment that local knowledge was at the heart of their success and needed to be relied upon to inform specific tactical decisions.

The other two layers of influence in the CDF decision making process were the president and government—exiled in Conakry May 1997–February 1998—and ECOMOG. The president remained the technical commander in chief and held the post of defense minister, although his involvement was by necessity truncated by his physical distance from events. ECOMOG, however, retained the overall command and the CDF were at all times, in theory, to follow the lead of the ECOMOG force. But whereas the president maintained closer links to Base Zero where his deputy Hinga Norman was based, and where operations focused directly on regaining key territories in the south and southeast to pave the way for an overthrow of the junta, ECOMOG maintained stronger links with Base One and the struggle against the rebels in the border region. But in both cases individual fighting units maintained a high degree of operational freedom. As a result, the actual command structure in the CDF likely came to look something like figure 6.4.

But even this already multifacetted structure was further complicated by the overlapping identities of some of the key contributing groups. Many of the advisers, such as Moinina Fofana and Allieu Kondewa, also held other positions, for example as war council members or even commanders. This meant that much of the decision making was pushed downward in the command chain, and some sideways, particularly tactical decisions.

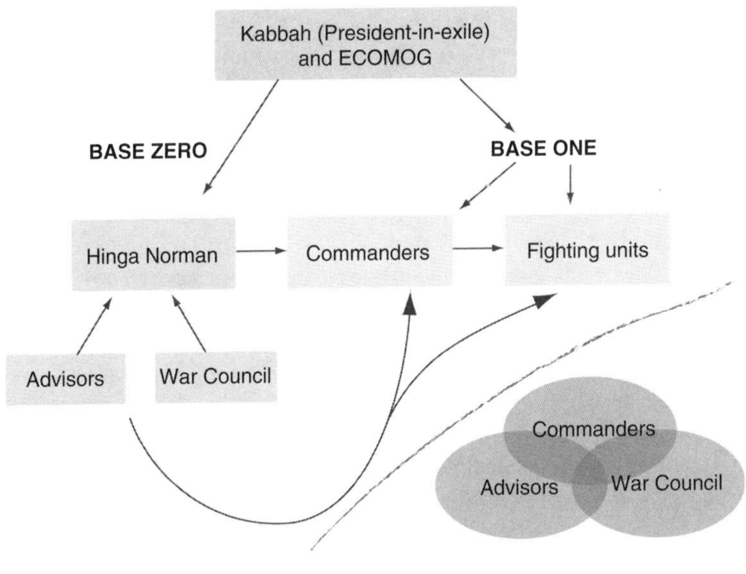

Figure 6.4 CDF command structure

Even for those operations that were centrally directed from Base Zero, plans often changed between Base Zero and the area of operations. An example of this tactical leeway of commanders could be seen during the famous attack on Koribundu during the later stages of Operation Black December in February 1998.[57] The commanders charged with securing the Bo-Koribundu highway and capturing Koribundu, appear to have formulated the specifics of their tactics only once they had left Base Zero. Originally, the Kamajors were sent by the War Council to the Koribundu/Bo area to attack the AFRC/RUF. To prevent junta reinforcements from reaching Koribundu from Bo (as had happened on previous attempts to take Koribundu) the "Death Squad" (a unit of 42 men led by a senior commander, Bobor Tucker) was sent to hold the Bo-Koribundu highway.[58] Another two commanders were sent with their units to take Koribundu. Yet all three commanders and their troops met in Kpetewoma, west of Koribundu, to plan the attack. At this planning meeting, the troops were reorganized and the Death Squad was reinforced by another 40 men originally intended for the Koribundu attack. They also decided to see through their mission by launching simultaneous attacks on Koribundu from the Sumbuya-Koribundu highway and from Blama. The attack was launched on Friday, February 13, 1998, 2pm, and by Saturday afternoon Kamajors entered and proceeded to capture Koribundu.

But despite the leeway clearly given in the Koribundu case, the planning of the deployment was expedient and maintained a recognizable chain of command. This was largely because of Koribundu's proximity to Base Zero, which made communication easier and allowed for a greater degree of involvement by the Base Zero command. In many other cases this chain was far less rigid and there was a significant element of commander inflation in the CDF. In some ways the Koribundu attack was therefore not a representative case, yet hinting at the increasing diffusion of central command structures the further away from Base Zero operations took place. Most operations that took place further away from the Kamajor "heartland"—close to either Base Zero or its key strategic targets Bo and Koribundu—were controlled to a far lesser degree by the central command.[59] Such control was impractical as communication was cumbersome and successful operations often depended on swift action and local knowledge, which were both facilitated by independent initiatives by local commanders. Although some appointments and promotions up the ranks were centrally administered most were handled locally, creating layers and strands of allegiance and command rarely effectively controlled by the leadership at Base Zero.

In theory, the fighting ranks of the CDF were organized into battalions, companies, platoons, and sections. In reality, these were as much "titles" as were many of the ranks and signified only a group of fighters, the relative power of which would differ depending on the area and the local chain of command. The numbers of these layers of groups varied, but could reach more than 500 where commands were combined which often happened in chiefdoms to maximize the available capacity.[60] However, in practice, units were rarely referred to in these terms and their numbers and organization were more ad hoc.

Overall the CDF was characterized by a strict organizing *principle* rather than institutional structure, whose application was nonetheless, as with most guerrilla forces, both compromised by circumstances and allowed flexibility to effectively execute strategy. Some of the overlapping layers of command can be explained by the recruitment process and, in particular, the central role of initiation that created bonds stronger and more immediate than what could be commanded by the official organizational structure. We shall revisit this in chapters 7 and 8.

LOGISTICS AND TRAINING—A RESOURCE GAP

The organization of logistics and military training for the CDF also betrays the localized nature of the organization. Training took place at Base Zero and Base One, as well as at recruitment centers at District level, but was generally rudimentary and not provided for all new recruits. Where it did take place, training would be led by ex-servicemen, and would trickle down as new recruits rose in the ranks and acquired skills and experience.[61] At any given time up to 5,000 recruits could allegedly receive training at Base Zero, where special training grounds were set up in the school field at Talia.[62] Most training would, however, be given "on the job" by more experienced fighters to new recruits. Some such training had also been given by EO to the Kamajors they worked with while contracted by the NPRC government.[63]

The weapons available to the CDF varied from place to place and at the different stages of the war. Original stocks of shotguns, cutlasses, and sticks had been the arsenal with which the local defense units had emerged, and in many cases remained their main type of weaponry throughout the war. But supplies of more advanced weapons and weapons captured from rebels added to this arsenal with AK-47s, lightweight machineguns, rocket propelled grenades (RPGs) and mortars. In the early days some of the defense units had been supplied by the NPRC government, and some of these weapons remained in circulation. New supplies came from ECOMOG, primarily in Liberia where some CDF military training had also taken place under ECOMOG supervision, and where possible from the government in exile. Weapons intended in part for the CDF were famously intercepted by ECOMOG in Sierra Leone at Lungi airport in February 1998 in what came to be known as the Sandline affair. 35 tons of weapons were flown in as part of an agreement between the Sierra Leone government in exile and private military contractor Sandline to aid the overthrow of the AFRC junta. But the shipment was in breach of the UN arms embargo on Sierra Leone and was impounded by ECOMOG. A small number of these weapons (250 rifles, 10 machine guns, and 100,000 rounds of ammunition) were, however, allegedly passed on by ECOMOG Commander Maxwell Khobe to the CDF.[64]

Communications between CDF forces and units and the central command was conducted through both the use of modern communications technology, such as radio and a satellite phone, and traditional methods relying on messengers traveling across the country on foot. Through the support of EO and ECOMOG a number of helicopters were used primarily by the CDF leadership to travel in and out of the country as well as between CDF strongholds.

The majority of CDF fighters were, however, forced to rely on limited weapons supplies and communications technology. A typical CDF outfit would travel through the forest by foot, with very limited supplies and more often than not simple shotguns and cutlasses. Overall, the arms used by the CDF were basic, and their military organization largely dependent on the motivation of its fighters and their ability to use to significant effect the most basic of military means. The spiritual dimension to their fighting therefore became an even greater resource to fill some of the material gaps in their supplies and training.

Internal Discipline and Codes of Conduct

Just like flows of logistical support suggest a more localized power base for the civil defense, the structures for maintaining internal discipline point to codes of conduct largely maintained by patrons with respect to their clients in the defense units. However, at the same time, national structures were created in parallel and sometimes overlapping with the reality on the ground.

In addition to the military offices in the CDF, civilian positions such as welfare officers and local administrative officers were set up at chiefdom level to liaise with the civilian population. After Kabbah was reinstated in 1998 the CDF's position was formalized and District level administrators introduced who reported to the Paramount Chiefs. This was in part to restore the reputation of the CDF that had suffered during the particularly brutal campaign against the AFRC/RUF in late 1997 and early 1998—despite the introduction of the War Council—but also to attempt to reign in some of the more subversive elements within the CDF ranks and reinforce central authority.

At Base Zero, the War Council had been in charge of disciplining fighters, although this appears to have taken place only occasionally. Reports of misconduct during operations would be presented to the Council by commanders or, if applicable, the affected parties. It was, however, generally at the discretion of commanders to bring disciplinary matters to the attention of the War Council, and as a result most incidents of ill discipline went unreported.[65] After the restoration of Kabbah's government and the relocation of many CDF officials from Base Zero to Bo, this responsibility officially came to rest with the District Administrators.

The CDF took pride in its code of conduct that was largely tied to the Kamajor rules for behavior and taboos that came out of their initiation. The rules, which included not targeting civilians and (on some accounts) respecting civilian property and refraining from looting, were designed to maintain the respect and support of the local communities to whose needs the Kamajors (and in extension CDF) claimed to be responding.[66] In this sense they grew directly out of the very raison d'être of the movement. But although the law preventing the harming of civilians was regularly quoted by more senior ex-combatants, the rules on looting appear to have been less clear among the ranks.[67] Nonetheless, the code of conduct was an integral part of the self-perception of the fighters who saw themselves as the antithesis of the rebels and "sobels" plaguing the country and its population. The respect for the "people" and the national fabric therefore became an organizing principle of the movement, even though it was not always followed to the letter. Ferme and Hoffman have argued that the codes in part grew out of the community roots of the movement but in part also responded to a more international discourse on human rights and the laws of war, disseminated through international analysis on for example BBC's Focus on Africa broadcasts and through documents produced and distributed by international organizations working in the region. They explain how,

> [i]n addition to reading aloud to their men from military manuals, commanders or their literate secretaries sometimes read from reports and guidelines by Amnesty International and other human-rights organizations as part of their training programs—for the express reason that a legitimate fighting force needed to be conversant in the laws of war.[68]

They continue to quote from a CDF report from 1999 by the CDF-restructuring committee entitled "Recommended Values and Standards," where the organization's relationship to international law is clarified:

> One of our ultimate desires in taking arms was to restore RULE of LAW of this country. It is therefore obligatory that as peace and tranquillity gain

> momentum in the country, every CDF personnel is subject to the Civil Law wherever he finds himself serving, and has a duty to uphold it. In that respect, CDF personnel are not different from other citizens, and therefore shall be punished for all civil and criminal offences as stipulated in the constitution of this country... In addition, we have resolved in our Restructuring process that when deployed on operations, CDF are subjected to the laws of armed conflict and to the laws of the locality in which they find themselves deployed.[69]

These guidelines were, however, formalized with reference to international standards, as above, only after the restoration of the Kabbah government in early 1998. Although it is a pertinent illustration of the ability of armed actors in remote corners of the world to access and make use of global discourse to their advantage, the strongest sense of behavioral restraint continued to come from the taboos of the Kamajor society.

The guidelines, and in particular the treatment of civilians and enemy prisoners, also appear to have been highly dependent on the given strategic context. In some cases enemy prisoners were not only kept alive but integrated into the fighting group by the capturing commander, whereas in other instances they were summarily executed.[70] Civilians could be treated in an equally varied way, and particularly during Operation Black December many civilians were targeted and killed on the assumption (often unsubstantiated) that they were AFRC/RUF collaborators. During Black December, fighters were specifically told to kill all collaborators before deploying for the operation, and were led to believe that civilians in the areas concerned that were not collaborators would have left on the instruction of the CDF over radio broadcasts. The implication of this was a blanket sanction on summary executions.[71] This seemingly contradictory approach to targeting civilians is no doubt in part a reflection of the perceived stakes during this particularly intense period of fighting.

Another factor was the level of commitment by a fighter who had only been through brief initiation during the mass recruitment drives, particularly in the later days of the war. One senior commander commented on the importance of the Kamajor society and "true" initiation that this is what allowed a commander to command his men. The sense of loyalty and purpose that stemmed from the initiation experience provided the commander with the power and authority to effectively direct his fighters.[72] The importance of initiation and the implications it had for command and control will be further explored in the next chapter, but it is important to note how the growth of the CDF as an organization (albeit a nebulous one) both worked in its favor as it increased its ability to mount successful attacks on AFRC and RUF positions, but also eroded some of the discipline that had given the movement its original strength.

THE WAR AND THE KAMAJOR "HABITUS"

The habitus of the Kamajors was shaped in part by the social and political reality of Sierra Leone as it had come to exist under the rule of Siaka Stevens and Joseph Momoh. It incorporated both the older traditions of patron-client relationships awarding power with responsibility to chiefs and leaders, and the corruption of centralized state-power that had skewed these relationships in favor of an increasingly small number of people. This contributed to a situation where armed insurgency could take hold, as well as to the difficulty faced by the central government in trying to resist the armed challenge. It was because of this difficulty that civil defense became necessary.

The CDF was an organic response to the rebel war that became co-opted by the State whose armed forces betrayed it first through quiet enemy collaboration and later through direct confrontation in the 1997 coup. The CDF would, however, have been unlikely to emerge in the form it did without the legwork of the local defense units and hunter militias that emerged at a much earlier stage. The overlapping yet distinct identities of the CDF and the societies that made up its forces, like the Kamajors, contributed both to the initial strength of the CDF and its disciplinary problems in the later years. However, the centrality of society practice—and particularly the Kamajor society as the most numerous as well as that to which the CDF leadership belonged—makes it crucial to understand the dynamics of that practice.

The examination of the organizational structure of the CDF turns up regular references to the initiation practice and the role of the taboos instilled as a result. The centrality of the initiators to the organization as a whole, and the combination of legal and normative frameworks with an international tone and locally anchored behavioral rules and taboos, illustrate the multifacetted nature of the movement. It also echoes the similar combinations of tradition and contemporary practice and concerns we encountered in chapters 2 and 3. The porous and flexible chain of command that in part resulted from practical concerns with overcoming physical distance and applying local expertise and capacity was further enforced by the importance of allegiances formed through initiation as well as traditional patron-client relationships. As a result, in the symbiotic relationship between the CDF and the societies that provided most its fighters, the societies and local organization were the key parties as they were the loci for both action and much decision making. They were also where magical practice was anchored: The habitus of the Kamajors was shaped as much by their own traditions as it was by the war, and crucially by the local cosmology. These were the reference points that could be consulted to design an appropriate civil defense mechanism, and the reason why magic became a central part of it. This is why in turning to explore the role of magic in the Kamajor society specifically we must look first at their cosmology. It will allow us to identify the nature of the connections between these realms and their relative contribution to the overall dynamics and practices of the movement.

CHAPTER 7

SPIRITUAL CONTEXT AS HABITUS AND CAPITAL

To contextualize the particular case of the Kamajors, we must look more closely at the particular beliefs and practices that constituted the cosmology and spiritual habitus within which the armed groups emerged.

Sierra Leone is a multireligious society, known for the ease with which Muslims and Christians have lived, and continue to live, side by side. Although most of the population is Muslim, traditional beliefs coexist with this monotheism and often unite adherents to the different belief systems.[1] A "secret society" tradition crosses religious boundaries, and draws heavily on traditional beliefs and rituals. The Poro and Bondo societies for men and women respectively, are the most widespread and function both as an educational framework to induct youth into the community's way of life and adulthood, and as networks for continued information and knowledge exchange. The beliefs that underlie much of the society practices are largely indigenous, even though as with most indigenous beliefs across the continent, they have become closely intertwined with Abrahamic traditions. In function, they are remarkably similar to many other human associations found around the world. Similarly, much of the specifics of the ritual practice and magic used in Sierra Leone is on closer inspection also familiar from far-away lands, suggesting perhaps both a continuity of human need and a heritage of a centuries old process of globalization.

The Kamajors are a Mende group, and the focus here will therefore be on Mende cosmology, although many of the characteristics cross ethnic communities and boundaries.[2]

MENDE COSMOLOGY AND SPIRITUAL PRACTICE

The "supralevel" of Mende cosmology includes a creation story that sets the scene for all subsequent interaction between the spiritual and material world. According to this story, all life and activity derives from the Supreme God, *Ngewo*. *Ngewo* not only created the universe but also invested it with nonmaterial power, which has been described as "an influence which is not directly visible to the naked eye but manifests itself in various ways and on special occasions in human beings and animals, and even in natural phenomena such as lightning, waterfalls, and mountains."[3] This power left behind by *Ngewo* is referred to as *hale*, which sometimes gets translated as "medicine" although it tends to signify something much wider. It is that (material objects—animate or inanimate) which can be used to secure ends through nonmaterial means by virtue of its power. But after creation, *Ngewo* withdrew from his work and can only be approached indirectly through spirits or ancestors who can mediate requests. Although the spiritual power left behind by *Ngewo* is available to anyone, the skills required to make use of it safely and responsibly take long apprenticeship and practice.[4]

Muslim and Christian beliefs sit adjacent to this approach to a spirit world and spirit power and get woven into the traditional story, allowing adherents to subscribe to both. In many cases, as we shall see, traditional structures draw directly on Muslim or Christian doctrine as indispensable parts to its practice. As Islam is the predominant Abrahamic tradition in the Mende areas of Sierra Leone, these have become more widely incorporated also in traditional practices.

The spiritual world in Mende society is engaged on several levels of practice. These levels of spiritual practitioners correspond roughly to the extent of their legitimate authority: the secret societies already mentioned, medicine men, mori magic, and witchcraft.

Secret Societies

Secret societies are a central part of spiritual practice and serve broad societal functions. They are widespread as well as socially sanctioned and legitimate. The practices of officials and senior members of the societies are approved of and seen as beneficial to the wider community. The societies are initiation societies and date back at least to the eighteenth century. Men and women are separated into the two gender specific societies, Poro for men and Bondo for women, and are invited to initiation at puberty.[5] Initiation generally takes places in the bush or at the edge of the town or village as secrecy is of primary importance. But despite the importance placed on secrecy, the youth's status as initiates is clearly displayed to the rest of the community. Although initiates are allowed into the villages during the day, they must return to the bush at night, and smear themselves with white clay whilst in the village so they can easily be recognized.[6] In this sense, it is the substance of the knowledge imparted to initiates that must remain secret rather than their identity as members, which on the contrary is a significant marker of social status. What they learn in the bush remains a secret to noninitiates and thus a source of power to those who have become members.

The role of the societies is primarily practical in that it divests its young initiates with knowledge useful to everyday life and adulthood. But it also creates a social network and hierarchy that can be called upon in times of crisis or for advice and adjudication. The societies serve as focal points for community organization and complement other power structures such as chiefs and national administration. In some ways, the societies act as an educational system with a strong element of the "old boys" network, although the same applies to the female Bondo society. The men and women in charge of the societies become powerful and important figures in local affairs and are often consulted in all manners of affairs relating to social and political needs. Caroline Bledsoe and Kenneth Robey have commented on the role of the societies as [managing] "political power through ritual authority based ideologically on hidden and allegedly dangerous and powerful knowledge that is inaccessible to noninitiates as well as to low-ranked members of the societies."[7] Mariane Ferme argues along similar lines that "[i]nitiation is a violent, painful experience, one in which big women as much as big men wield the knife and in which the urgency of protecting secrets and bodily boundaries is inscribed in the initiate's flesh."[8] That said, the extent of the secrecy is limited in that, at least with the main two societies—Poro and Bondo—a large number of community members partake in initiation. Although membership is in part limited by cost—there is a fee for initiation—it nonetheless attracts significant numbers of recruits. There is also a strong sense of tradition and families of initiates tend to send their children for the same experience if means allow.

Despite the limitations on the anonymity of initiates, there remains a strong wall of secrecy between the two main gender specific societies. For example, although it is widely known that girls go through circumcision as part of their initiation, the details surrounding the ritual is obscure, and most men are unaware of the specifics.[9]

Paul Richards has written about the rituals associated with initiation as a Durkheimian performance intended to create the bonds necessary to hold the group together. Where Ferme sees boundaries inscribed in the flesh, Richards adds "social glue" that, as a result of these boundaries and the sharing in their inscription, binds the members together. He compares the functions of these rituals of initiation with Roman and medieval European sodalities that through rites created societies of craftsmen and traders that reached beyond the purely material and incorporated a spiritual dimension. Like Durkheim, he is reluctant to ascribe importance to the beliefs themselves and rather sees the act of ritual as the decisive formative experience: "Feelings have to be stirred before they can be shaped. Ritual conveys to a group a sense of its power as a collectivity... Ideas about spirits, ancestors and gods are secondary."[10] Although discarding the relevance of the beliefs altogether may be going too far—as shall be argued below—the importance of translating the beliefs into action, and particularly collective action, lies at the core of the secret societies.

This framework of secrecy, bonding, and loyalty in varying proportion, is familiar from activity around the globe. It is not a Sierra Leonean or even West African phenomenon. Secret societies and initiation rituals can be found around the world. In the West monastic orders were the early bastions of fraternity, pledged loyalty, and common purpose. More spiritually eclectic societies such as the Freemasons and Rosicrucians followed, with greater emphasis on initiation and mystical secret knowledge in addition to the format of a bounded society and strict loyalty.

There has also been significant cross-fertilization between societies and different parts of the world. Europe has exported its Freemasonry and Rosicrucianism to all corners of the world, and West African style initiation societies can be found across the Atlantic.[11] The smooth translation of society practice from one part of the world to another has undoubtedly been aided by their similar functions even if they vary in form. On the one hand secrecy, and particularly layers of secrecy made accessible through progressive steps of initiation, provides a means of exercising control. Withholding knowledge from lower level initiates as well as the community at large, while hinting at or clearly expressing its existence allows for the exercise of power over those who are seeking the knowledge on offer. This can mean financial rewards in return for initiation, or other favors or designated behavior. But on the other hand, the knowledge distributed serves a purpose of its own. In essence, the levels of secrecy open up for a reward system that offers benefits for both those divulging and those seeking knowledge. In this sense the secret societies are also a means of spreading knowledge and building networks that are beneficial for their members. And in some cases they can be more akin to educational institutions, with entry requirements and tuition.

In the secret societies of Mende Sierra Leone we are therefore faced, not with a peculiarly local phenomenon, but with a local variation on a virtually global theme where different forms of human association help structure interaction and form norms in a given community. And in the case of the Poro and Bondo, secrecy—although in principle central to the societies—is rather less important than the loyalty and bonds created by the shared experience of initiation, and in extension social order. Because of the centrality of traditional beliefs in the habitus, the secret societies take on the specific form that they do, but their functions are nonetheless more universally recognizable.

But just as there are more or less accepted secret societies and associations elsewhere in the world, there are secret societies in Sierra Leone that do not fall under social control and sit outside social sanction.[12] These are the societies used not for the benefit of the society as a whole but for individual gain. Most notorious in the Sierra Leonean context are the Leopard and Alligator societies. Notorious primarily because of their alleged widespread use of human sacrifice, these societies became the focus of colonial administration attention with a number of trials of members before a special commission court in the early twentieth century.[13] The societies, or variations of them, still exist in the region and have been documented particularly in Liberia.[14] During the war, echoes of the more extreme ritualistic practices of sacrifice practiced by the societies emerged through cannibalism and ritual killings. They do, however, place much greater emphasis on secrecy—both as regards membership and practice—largely as a result of the negative response to their activities by the wider population. But in spite of their more extreme form, and generally less common practice, the spiritual basis remains the same as that of Poro and Bondo. And, as explored in chapter two, it is the process and use to which the power is put that separates them from the legitimate societies, rather than the practices or beliefs themselves.

Medicine Men and Mori Magic

Medicine men and practitioners of magic and divination float on the boundaries of the socially sanctioned and the more suspicious private practice beyond communal control. Although they are at times employed for communal purposes, generally their services are reserved for private ends.[15] Medicines are compounds of natural ingredients and are used for ritual cleansing or to cure illness.[16] They can be ingested or worn depending on the purpose for which they are being dispensed.[17]

The medicines themselves are used both by the initiation societies and individual practitioners, and again, it is the *use* to which the knowledge and craft is put more than the power source itself that is the basis of legitimacy. That said, not all private practice is deemed illegitimate, although private ends that cannot be immediately connected with a societal good are more open to interpretations and accusations of witchcraft, or, destructive intent. This is reminiscent of the cunning folk familiar from an older European setting as well as many parts of Africa, explored in chapter two. Balancing on the edge of the sanctioned these practitioners of herbal medicine, with a dose of spiritual summons, become defined in large part by their intent and the direction of their craft. These practitioners are, nevertheless, consulted by a wide range of individuals from all walks of life and act both as standard health professionals and miracle makers.

Another variant of individually practiced magic is that of the *morimen*. *Mori* practitioners are Islamic and Arabic specialists who use their knowledge of written Arabic and the Qu'ran for ritual and magic purposes. Their power derives from the correct application of a combination of written Qu'ranic passages, blessings, prayer, time, and space. The underlying belief is in the power of the written word, in particular the divine words of the Qu'ran, and builds on a long heritage of Islamic mysticism and magic.[18] The use of written Qu'ranic verses for magical purposes goes back to the early days of Islam and has been practiced throughout the Islamic world. It found particular resonance amongst Shia sects and the Sufi traditions, which were also those that spread most effectively down through the Sahara to West Africa along the ancient trade routes.

There are a number of ways in which the power of the Qu'ran and its written words can be used by the specialists: Qu'ranic texts can be written on note paper, folded and tied up or placed in a pouch as a charm. It can also be written on a board (*wala*) with black ink or a piece of burnt wood from the *luba* tree, then washed off to create a dark liquid called *nesi*. The liquid into which the words have been dissolved retains their magical properties and can then be drunk or smeared on the body for magical effect.[19] The Qur'an can also be used for divination (*fal*) and guidance (*istikhara*), where the specialist will ask questions and turn to the Qur'an for answers. By interpreting the texts on the pages opened, according to specific rules and keys, the specialist will divine the future or provide guidance in respect to specific questions.[20] The same principle has been applied for centuries across the Islamic world and has left behind an extensive heritage of specialist tools such as copies of the Qur'an marked for *fal* and *istikhara*. Pouches and medallions containing prayers can be found in market stalls in Istanbul as well as Khartoum, and seen on cattle herders and nomads across the Sahel.

The morimen can also draw on the help of spirits—the *jinai*—to empower their art. The *jinai* are powerful spirits that can make special demands of God, by exerting pressure on ancestors to bring their request before God. Again, they can be good or bad, and although they are traditional spirits, they are allegedly generally willing to assist Muslims well versed in the Qur'an.[21]

As with the initiation societies, secrecy is central to the power of the moriman and his craft. There are several layers to the moriman's magic that add to this secrecy: he must be literate in Arabic, in itself a time-consuming undertaking added to by the reluctance of many morimen to pass on too much knowledge to their apprentices lest they become rivals; he must understand which are the relevant texts in the Qur'an and what their meaning(s) are; which verses have magical powers and in which specific instances they can be applied; knowledge of prayers, and the combination of text, prayer and blessing required for different rituals to be effective.[22] When dispensed, the magically imbued substances (like written notes or *nesi*) are kept hidden from view, keeping them secret not only to their bearer but also to the outside world. Once mastered, the moriman can use his power to help clients (which are generally paying customers) to do everything from divine the future to cure illness and pursue material goals.[23] Because of the wide reach of their power, morimen are often employed as ritual advisors to Chiefs and Big Men, but are also consulted more widely by members of the community who can afford their services.

But just as medicine men can be employed for societal purposes, mori magic is not an entirely privatized realm and features prominently in initiation societies. For example, the masks used in Poro and Bondo ceremonies are often inscribed with Qur'anic verses and mori magic as are the rituals performed.[24]

SECRECY, THE HIDDEN AND POWER

A common theme that runs through Mende cosmology, and to a degree African traditional belief more generally, is the centrality of secrecy and the hidden to the acquisition of power. For the Mende, the secret societies, herbalist and moriman-practice, all value and pay tribute to the hidden and secret. As alluded to above, the hidden and secret allows for a greater sense of mystery but also community among those party to the secret knowledge. In societies it promotes loyalty and community, and for practitioners of magic like herbalists and morimen it guards their professional space.

In many discussions on the nature of African politics and its interplay with the spiritual, reference is made to the "politics of the belly."[25] The idiom of ingestion as a reference to the acquisition of power is a common one and connects the material with the spiritual while also emphasizing the impact of the unseen. Spiritually powerful men and women are often said to have a substance—a physical object—in their stomachs that provides them with power. Again, it is an attribute that cannot be detected from external physical appearance but that nonetheless can impact powerfully on individuals and events. When the Revolutionary United Front (RUF) rebels invaded Sierra Leone from neighboring Liberia, they had amongst them mercenaries from Burkina Faso. These men were rumored to have strong spiritual power, and Kamajors who caught and killed some of them explained how they had found charms inside their bodies.[26] Whether they had in fact ingested such charms and amulets as a result of their own beliefs in their power, or whether this was detected by their captors because of their expectations, or not at all, is hard to say. But the idea of a material, physical substance providing power through ingestion, has given rise also to less palatable practices such as ritual killings and cannibalism. The most notorious practice is the extraction and ingestion of an enemy's heart and liver to absorb his strength and power. And the use of body parts in medicines and charms come again across the continent.[27]

But there are ambiguous aspects also of the relationship with secrecy and the hidden. Not all charms are ingested, and some are even left in plain view. As such they send the signal that there is magic and power at work, although the nature and extent of that power remains hidden in the wrapped up charm or amulet. Similarly, the new trainees for the Poro and Bondo societies signal their status by smearing white clay on their bodies—remaining in part hidden in the society bush, but also seen as undergoing the process of initiation. This ambiguous interplay of the seen and the unseen allows for the secrecy to have more far reaching consequences as it becomes part of the symbolism of everyday life (and more extreme situations such as war). In this sense, individuals in part realize power through a display of signs that this power is possessed. This is not dissimilar to military insignia or profession-specific uniforms that signal a particular type of power, be it that of a legal enforcer, doctor or even more subtle insignia such as university signet rings or Masonic pins. By conveying membership to a particular group or corps and a hierarchical position within it, power is also communicated and can in some cases have returns that reinforce that suggested power in real terms. This generally happens when the symbols are interpreted by others and their behavior is altered because of this interpretation.

In Bourdieu's language, this is a part of the habitus that impacts not only on perception and interpretation but also on bodily posture—the *bodily hexis*.[28] For this reason, interaction between people who share in the habitus that embodies belief in the supernatural will be guided by interpretations of behavior and appearance that carries significance only to those who share in that habitus. During war, this type of display tends to be more obvious and specifically intended to convey and exert power. The particular impact of signs and symbols of secret power is dependent—as much as other symbols that lack the element of secrecy—on an appreciation of that power by those who see them. In Sierra Leone where these signs are generally known, the effect also becomes more tangible.

Although the particular form of the initiation societies in Sierra Leone and the rituals of spiritual practitioners are the products of the local communities that have given them their specific form over the years, they are not altogether unique. On the one hand, they draw on practices and beliefs that have filtered through to the region over years of exchange with traders and visitors from far afield. Both Islam and Christianity have been introduced to the region and incorporated in the spiritual habitus of its people, providing a tangible link to an outside world and a wider context. At the same time, beliefs and practice of external origin have almost exclusively been blended with the local cosmology and incorporated in it. The practices that result appeal to human needs of social structure and organization encountered across the world. But at the same time, it is the specific habitus that informs perceptions and interpretations that give rise to particular types of interaction. This combination of the universal and the local is what we see also in the wartime practices of the Kamajors and the CDF. And through the examination of the local and specific vis-à-vis the habitus and field within which they exist, we find a logic that stretches much further.

The spiritual dimension to the Kamajor society reflects and draws on the traditions outlined here: in the initiation rituals performed on recruits, in the rules and immunization rituals performed on fighters, and in the secret nature of the society's knowledge and practices. But just like the Mende spiritual habitus has evolved and absorbed new ideas and practices over the years, so the Kamajors, while drawing on practices with long traditions, familiar to recruits, also reinterpreted and shaped these traditions for the specific purpose of war and the constraints and opportunities it affords.

CHAPTER 8

Creating Magic Soldiers: Kamajor Mobilization and Initiation

If you lose trustworthiness, the lore will be ruined.[1]

The exact chronology and historical narrative of the Kamajors' involvement in the war is hard to establish in detail, beyond some well-documented key events. In chapter 5 we traced the evolution of the civil defense units, and the Kamajor society, to a national Civil Defence Force (CDF) only to identify the discrepancies between the official structure of the organization—or at least the way in which it tends to be understood when taken at face value—and its actual working on the ground. Similarly, the perception of the Kamajors by others, their self-perception, and actual events do not always add up. But the power of the Kamajor society and fighting force, in so far as they were conceptualized—by themselves and others—as a military force with supernatural powers, is in a way separate from an accurate historical account of the movement. The perception shapes and guides action and as such impacts on success. It is important to bear in mind this slight disconnect between the accurate historical narrative and the individual narratives that combined to influence action and interpretations of events.

This chapter explores the role of initiation and ritual in creating the force and constituting its fighters as powerful warriors—in their minds as well as in the minds of the wider population. First it will consider the idea of the "Kamajor" and its mythical relevance in the Mende southeast, and, its subsequent translation into a real fighting force. Secondly, it will look more closely at the motivation of recruits for joining the society. Finally, we will explore the role of the initiation itself, in order to explain the glue that both attracted recruits and held together the movement.

THE MYTHICAL HUNTER AND THE KAMAJOR TRADITION

"Kamajor" is the Mende word for hunter. In the colorful tradition of the Mende language it reflects multiple meanings, and as a result the academic literature tends to translate it with some variation. Patrick Muana suggests it derives from the words *kama* and *joi* translating literally as "a past master at doing mysterious things," and others suggest as an alternative "masters of marvel."[2] The general meaning, however, is simply hunter, although with the implication of mystical powers. The Kamajors were traditionally Mende hunters, in southeastern Sierra Leone. They were few in number and charged with the protection of the community from primarily wild animals, but occasionally human enemies.

In his careful consideration of the Kamajors, Danny Hoffman traces the role of the hunter and the "mythico-poetic origins" of the Kamajor in an illuminating manner: He suggests the first Kamajors were wanderers in the bush, killing wild animals that were causing trouble for their village. Their prey would be offered in part to the local chief and in part sold. Because of their connection to the chief, he would be held responsible for the actions of the Kamajor—a connection that survived in the recruitment process during the war. Hunters who killed large animals were alleged to have magical powers that helped them in their work, provided alternatively by morimen or the animals themselves.[3] But there were only a few hunters that were also Kamajors. What distinguished the Kamajor from other hunters was his command of the bush paths, the gun, and the secrets and medicine of the forest.[4] Significantly, however, Hoffman notes that the Kamajor did not have a society but operated on his own after a long apprenticeship to acquire the skills and knowledge that constituted him as a Kamajor and not merely a hunter.

The powers and skills of the Kamajor also made him ideally suited to provide defense and war craft when called upon. The ability to track and kill large animals in the dense bush equated not only with physical perils but with spiritual danger, and set the Kamajor aside from other members of society, both as an awe-inspiring character and a somewhat unpredictable and potentially volatile force because of his specialist and secret knowledge. The power associated with knowledge and in particular secret knowledge, as seen in the earlier discussion of Mende cosmology, was recognized throughout Mende Sierra Leone and helped constitute the figure of the Kamajor as one of power.

The Kamajors in their guise as traditional hunters were also central to many narratives about Mende origin. They were the ones said to have spearheaded new settlements and, once established, ensured their defense and day to day running:

> Endowed with powers of divination, invincibility, and omnipotence, the Kamajoi embarks on a symbolic quest into the unknown. Alone or accompanied by a reputable "medicine man," he travels to distant lands, conquers either man or nature and starts off a settlement... The Kamajor all at once straddles and transcends this tripartite but intertwined Mende cosmological divides of man, forest/beast and spirit.[5]

The position occupied by the figure of the hunter in Mende imagination was therefore closely connected to power and violence, and the transition to an organized military force was a logical one in times of war. However, most men initiated into the Kamajor society during the war were not hunters, nor did they acquire hunting skills in the traditional sense through their initiation and training. The Kamajor society as it came to exist during the war drew on the image and myth of the Kamajor more than on any tangible tradition. Nonetheless, this image was the natural point of reference for the figure of a fighter imbued with magical powers and tasked to serve the community, and as a result the adoption of the name immediately connected the mandate of the force with a recognizable tradition, and their ability to carry out that mandate with the acknowledged powers of the traditional hunter figure.

The transformation of the hunters into an organized CDF—the way we understand Kamajors today—was a war time phenomenon, prompted by the insecurity of the local population introduced by the spread of RUF control and abuse in the area as well as corrupt government soldiers not only failing to protect civilians but actively engaging in their abuse. Community defense had always fallen within the remit of the specialized adult hunters and secret society members, and the utilization of this same group in a more formal defense capacity was a natural progression.

As explored in chapter 5, the secret societies hold a central position in Mende life, encapsulating the norms of the community and facilitating most activity within the community. Initiation provided the education for how to become a full community member and at the end gain access to that community. During the war, across the Mende south and southeast of the country, the Kamajor society became the prominent point of access for young men not only to the perceived powers and protection awarded to members but also to the prestige and position associated with the society tradition. Indeed, Hoffman has noted that it was "no coincidence that to become a

Kamajor one must be initiated, and that during the war to become a man one needed to become a *Kamajor* in much of the rural Mende-dominant region."[6] In this sense, the Kamajors were anchored in the community in more ways than through its defense—something that also becomes clear with the approach to recruitment at least in the initial stages.

To appreciate the circumstances under which initiations began and how hunter militias were transformed into a secret society with a strong supernatural dimension, it is helpful to consider the "creation" story or myth of the Kamajor society, which became the springboard for broader mobilization. Secondly, we need to examine the ways in which this supernatural dimension was incorporated in the society's practices and how it affected its ability to mobilize support and recruit members. This is best done through a closer look at the initiation practice itself and the effects it had on recruits.

The Men from Bonthe and the Genesis of the Kamajor Society—Creating Legitimacy

Although the Kamajor society grew out of an existing tradition and in part practice (of the specialized hunting society), most of the practices that came to underpin the society as it grew forth during the war were specifically tailored to the context of armed conflict. As a result most of the practices were derivatives of traditional rituals adapted to the circumstance of war, and designed specifically for the militia.

Accounts of the events that led the original group of traditional herbalists and Qur'anic scholars in Bonthe to come together and create specific initiation practices for Kamajors during the war reflect a sense of responsibility toward the community familiar from the original figure of the Kamajor hunter. Although the accounts show slight variations, some themes are recurrent: loss, of family in particular, at the hands of the rebels and an impetus to fight back prompted by encouragement from beyond the grave.

In one account of Allieu Kondewa's (alt. sp. Kundorwai) beginnings as a Kamajor initiator drawing on both of these themes, the force's origin is framed as a sacred mission, bestowed on Mende men:

> Following an RUF attack on a village in the Jong (Jange) Chiefdom, the rebels are reported to have massacred people in the village including a great "Kamajoi" and medicine man called Kposowai. His brother Kundorwai, is said to have been captured by the rebels, forced to carry looted goods and tied ("tabay") securely for the night whilst the rebels pitched camp. As he drifted to sleep in spite of his pains, Kundorwai is said to have had a vision of his brother who had been killed the day before. The ropes fell loose and

> the elder brother invested him with the authority to take the message to all able-bodied Mende men that the defence of their own lives, homes, wives and children was a sacred duty. To assist them in that task, Kposowai is said to have shown Kundorwai a secret concoction of herbs and instructed that a stringent initiation process should proceed the "washing" of the warriors in the herbs. This concoction would make them invincible in battle, impervious to bullets, and endow them with powers of clairvoyance if all taboos were kept. Kundorwai is said to have then slaughtered the RUF rebels, freed the other captives, and trekked several miles to a secret hiding place where he initiated the first set of men.[7]

A different account emphasized how Kondewa also lost his wife and some of his children in the rebel attack, further strengthening his resolve to fight the brutality of the RUF.[8] An account by another initiator of the impetus for his becoming an initiator shows a similar pattern. He lost his wife during a rebel attack, and was later visited by her spirit at night at which time she showed him rituals and ceremonies of power to aid the struggle against the rebels.[9] These experiences provided not only personal motivation for the individual herbalists and scholars to become initiators, but also gave legitimacy to their roles and powers.[10]

The practical creation of the initial initiation cadre brought together experiences like these. The onslaught of rebel violence had brought rumors that some of the RUF rebels had come from Burkina Faso and had magic that made them immune to bullets. As a response to this threat that was perceived as not merely physical, traditional herbalists and Qur'anic scholars came together to construct a defense capable of resisting the Burkinabe witchcraft. Having already felt a need to fight back (particularly after becoming victims of rebel violence) the alleged witchcraft of the RUF provided an immediate role for herbalists and morimen in the defense effort. Together they designed rituals for initiation that would allow initiates to not only be immune to bullets but to become invisible, walk through walls, and command material objects.[11] They initiated the first group of Kamajors (in their name-brand guise) in 1995 in Kale, Bonthe district.

There are alternative (but generally complementary) narratives of the start of initiation and the birth of the Kamajor society as a magically empowered fighting force. One such narrative, as seen above, focuses on the person of traditional healer and herbalist Allieu Kondewa, and names him as the central figure who initiated the first 37 Kamajors in Kale, Bonthe district.[12] This falls in line with Kondewa's subsequent position as the High Priest of the society, but other accounts have suggested a flatter hierarchy and a collegial coming together of a core group of initiators. Regardless of the exact nature of the hierarchical relationship of the initiators, or lack thereof, it is clear that they worked together to design the rituals and practices. Some, including Kondewa, were illiterate and not proficient in the text-based magical practice of the Qur'anic scholars and as a result had to rely on others for some aspects of the initiation and ritual practice.[13]

The spread of initiation from Bonthe and the rapid growth of the Kamajor society was the result of a timely marriage of hunter militias already organized in the region and the initiation practices pioneered by the men from Bonthe. Rumors of the initiates and their success against rebels brought them to the attention both of the civil defense militia organized by Hinga Norman in his chiefdom Jiama Bongor in Bo district, and the NPRC in Freetown. Norman sent word to the Kamajors in Bonthe to come join his force and aid the SLA (under the NPRC) in their struggle against the rebels. The Kamajors, then under the military leadership of Mualimu Saddam Sheriff, came to join the militias, and the initiators sent Kamoh Lahai Bangura to establish a shrine for initiation in the area.[14] From this point on, the Kamajors as we have come to know them began to operate across the region.

Defense and Magical Powers—the Two Pillars of Mobilization

The idea of civil defense had already been anchored in the region through the mobilization of defense units in the various chiefdoms and towns. Eager to resist the threat posed by the rebels and the devastation they wreaked on the region, the concept was not a hard sell to most men of suitable age. The addition of the supernatural protection offered by the initiators and the Kamajor society, nevertheless added a significant impetus to sign up, and these two elements—civil defense and the supernatural powers gained through initiation—became the carrying pillars of the Kamajor society and civil defense movement.

Protection, Power, and Pride—Why Join the Kamajor Society?

In the Mende heartland few men did not join the Kamajors in one form or another. Whereas not all became fighters, many went through initiation and remained in their communities as a reserve force. One Kamajor commander listed four main reasons why men would join the society: for a sense of pride in helping to provide community defense; as a logical extension of the Mende tradition of waging war through specialized secret societies; for protection and to receive powers of immunity in case of attack; and because being a member of the society became "trendy" amongst young Mende men and as a result yielded social benefits such as respect and the attention of women—in effect it became a macho symbol.[15]

For most recruits their motivation to join and go through initiation was a mixture of these reasons, generally coupled with pressing security concerns. The status as pioneers and protectors of the Mende provided them with a deeply rooted legitimacy, but also power. As a consequence, the status of the society aided recruitment as it offered opportunities to access resources unavailable to nonmembers. These were partly material, as reflected by a young fighter's admission that "all the good things of the village go [to the society]."[16] But it also offered nonmaterial resources, such as power derived from societal standing and particularly access to spiritual power and the secret knowledge exclusive to initiates.

During the war more urgent needs for protection and survival dictated the choices of most men who joined the Kamajors. One male ex-combatant who started out with the Army originally demobilized in 1996 when the Kamajors had become a credible counter force against the RUF. But he took up arms again on the side of the Kamajors as President Kabbah was overthrown in 1997 and the national army was seen as colluding with the RUF:

> In 1997, when the soldiers overthrew Pa Kabbah they called upon all the ex-soldiers to join them, but we did not do it because they already mingled with the RUF. They were killing innocent lives and destroyed private properties so I did not join them. So we went into the bush to join the Kamajors.[17]

The theme of enlisting to protect one's home or community is echoed by a majority of Kamajor and CDF ex-combatants:

> [In] 1995 the war affected us and we had to move from the village to the displaced camp at RTI [Kenema]. There we were fighting for our survival and later the RUF drove us away from the camp and we went to Bandama to settle there for some time...Later they moved the displaced camp to Blama. So after all those hurdles I decided to join the CDF in 1996 to be able to fight for my village. [...] I decided out of my own free will [to join the CDF]. It was because we were tired of running from the rebels so we started to chase them from their territories.[18]

In a survey done in 2002 with ex-combatants, only nine percent of former CDF fighters said they were forced to join the force compared to 72 percent of ex-RUF.[19] Out of six lower ranking Kamajors interviewed by Krijn Peters for another project on ex-combatants only one said he reluctantly joined the force, but only in so far as he felt that being a civilian was no longer an alternative and he had to choose between the Kamajors, RUF, or SLA. He chose the Kamajors because he said "I wanted to defend my motherland and the soldiers have [*sic*] converted themselves to the RUF by then."[20] A 2005

Human Rights Watch report on the nature of young recruits and mercenaries in West African militias and rebel groups comes to the same conclusion, arguing that whereas most of the RUF or even SLA recruits found it hard to articulate the political objective of their armed group, the Kamajors (and CDF more broadly) presented a clear exception. In their case almost all said they joined to protect their communities from attack.[21]

Some initiates would spend a period of time as "inactive" members before joining the fighting ranks. One commander explained how he had been initiated in October 1996 but remained in Kenema as a civilian. After the AFRC took power, he heard they were hunting down SLPP members, and as a party member he fled to Liberia and joined the combat units there.[22] In other instances the roles would be reversed and men would join as fighters before going through initiation.[23] The reasons to join the Kamajor society all tied in with the habitus of the young men that made up the recruitment pool—for them the society offered a logical approach to dealing with the rebel threat, and one with benefits that could stretch beyond the immediate military challenge.

"All Able-Bodied Mende Men"—Who Joined the Kamajor Society?

To become a Kamajor, recruits traditionally had to be nominated by the community and vetted by the local chief—a practice that survived through the initial stages of the war. This process both assured the quality and dedication of recruits and ensured the close connection between the society and the community at large. Chiefdoms would be asked to contribute troops depending on their size, smaller chiefdoms contributing less and larger ones more. The chiefs were responsible for making sure that recruits were not or had not been members of the RUF or SLA (in the period after the coup), and had no criminal record. In some cases they also had to be "true" descendants of the village to ensure a wider net of accountability and responsibility.[24]

Like most secret societies in Sierra Leone, the Kamajors were traditionally gender specific in their recruitment. Although the vast majority of recruits were therefore male, there appear to have been some exceptions to this rule. Mazurana and Carlson in their study document the presence of female Kamajor fighters:

> Women and girls were fully initiated members of the CDF . . . Although the *Kamajors* were originally a male-only traditional hunting society, in response to the increased pressure from the RUF it became a self-defence force and enlisted women and girls beginning in the early 1990s and continued this practice throughout the war.[25]

Accounts of female combatants fighting alongside the initiated male Kamajors have also come from Kamajor commanders who explained the need to train the women helping out at the camps in combat to ensure their survival, but also to draw advantage from additional fighters. One commander in particular praised the female fighters in his unit who were fully integrated combatants. According to these accounts, the women were, however, not initiated—a practice reserved for men only.[26] Although there were some female fighters in the CDFs, and some most likely received some magical protection, it seems unlikely that they were in fact initiated society members as this would have been a major departure from the generally accepted boundaries of initiation practice and most likely counterproductive for an organization seeking community support.[27]

Interestingly, one of the main initiators in Bo, Mama Munda Fortune, was a woman. She was also the head of a fighting force known as the Kassela War Council. One Kamajor ex-combatant explained this apparent contradiction by her age and consequent lack of a "sexual persona," removing her from the realm of temptation otherwise associated with women, which is also the main reason quoted for not allowing them in the initiation society.[28] In addition to this, Mama Munda Fortune was not herself an initiate, although she had the power to initiate others. The taboo on women initiates was deeply engrained in the collective body of fighters, and it was generally believed that when dressed in their Kamajor clothes complete with amulets and magical protection, any woman that touched the fighter would bleed. However, in an opposite type of influence a Kamajor was also thought to loose his powers if he touched a woman before battle.[29]

As the war progressed, the process of recruitment was liberalized. Young men wishing to take up arms would join the fighters without community sanction and nonetheless be accepted for initiation. At times, men were allowed to join the CDF without being fully initiated Kamajors, protecting the integrity of the core of the society. One Kamajor ex-combatant who became close to the CDF leadership explained how he joined as a fighter but did not go for initiation until the High Priest requested he become an initiate. Initially he accepted only protection in the form of *nesi,* amulets, and charms, but eventually went through initiation.[30] A former RSLMF soldier that demobilized only to take up arms on the side of the CDF tells of how he was accepted as a fighter but could not be initiated as a Kamajor because he had been with the Sierra Leone army previously.[31] But more commonly recruits were accepted for initiation as long as they or their commander to be could pay the required fee.

The fee for initiation varied depending on the level of initiation and the period of the war. During the junta period when fighting intensified and recruitment increased, one ground commander said he paid Le 50,000 to be initiated.[32] Although the price for initiation—a direct heritage from the Poro initiation and somewhat of a safeguard against recruits lacking commitment—may have had honorable origins, the continued practice throughout the war and the price inflation was judged by some as a reproachable ploy by the High Priest and his initiation cadre for enrichment rather than a service to the communities. On the other hand, others have argued that the very fact that initiation required money allowed the Kamajor movement to operate more like a craftsmen's guild than a volunteer or conscription army, and the fighters to see themselves as specialists with particular rights and privileges.[33]

In spite of local rules against nominating them for initiation, fighters from the opposing factions were generally accepted into the CDF and put through initiation. Some changed their allegiances voluntarily whereas others were captured by Kamajors and joined as a result of their new situation. Overall, however, fighters moved with relative ease between fighting factions depending on the balance of power in a given area, but the move from rebel to CDF remained more common than the other way around.[34]

As the force expanded rapidly, and especially during AFRC rule when the Kamajors were effectively displaced from towns to Base Zero and Base One, the system of chiefdom nomination and control of recruits broke down. As a consequence, quality control of recruits became more lax and abuses increased. This liberalization of recruitment strategy has been pointed to as a key source of the breakdown of order that saw Kamajors and CDF resort to looting and other activities directly at odds with their mandate to protect the community.[35] The originally grass-roots and community driven force became progressively separated from the community it was intended to protect and less accountable to traditional authorities. This was particularly so where the Kamajors were deployed to fight in areas other than their own and were geographically separated from the immediate mechanisms for control ensured by the ties to a particular community.

In the later stages of the war, notably after the fall of the AFRC, initiation became even more separated from the original principles of the society. Hoffman notes that

> The market logic of initiations in this period were widely noted and instructive; for example, newly designed rituals which relieved the initiate of any need to observe taboos to maintain his bullet-proof status were more expensive than those which did not.[36]

The High Priest Kondewa led this development personally with specialized initiations, although he was repeatedly reigned in by the rest of the leadership, and in the end relieved of his duties as High Priest.[37]

The disregard for the traditional mechanisms for recruitment and initiation was also seen to compromise the spiritual power of the Kamajor fighters and in extension their general military power and legitimacy. This is made clear if one looks at the process of initiation, how it operated, and the effect it had on those who went through it.

Initiation—Creating the Group

Initiation was the crucial element to the creation of the Kamajor group. In the tradition of secret societies, the group was in part defined by the secret knowledge its members shared, knowledge that was divulged through the process of initiation. To become a Kamajor the recruit should in theory already be a secret society member, traditionally a Poro initiate.[38] Significantly, he would then also be an adult as the Poro initiation marks the passage from childhood to adulthood. In this way, the recruit would already be part of an exclusive group and would understand the philosophy of secrecy that runs through all aspects of Mende culture. The habitus would have been made more explicit, although all members of society shared in it in less pronounced and conscious ways by appreciating the power vested in society members of all kinds. In reality, however, Poro initiation became increasingly difficult during the war, and the pressing need for fighters allowed this rule to be compromised. As a result, the Kamajor initiation would become even more central to the identity of the recruit as it was his first (and often only) secret society experience as well as one that took place under the added pressures of armed conflict.

The initiation process is by definition secret and as such it is difficult to verify accounts of its constituent elements. Various accounts nonetheless converge on a basic description of initiation as provided below, albeit qualified and supplemented where necessary.

Initiation into the Kamajor group drew on the same spiritual tools as the Poro and Bondo societies and incorporated mori magic. A *kamoh* (initiator), with specialist knowledge of the Qur'an and mori magic initiated the new recruits. As with Poro initiation, the initiation ground was separate from the rest of the camp or training ground, and especially designated for the rituals of initiation, connecting this new practice with already established and accepted traditions. Descriptions of initiation reveal a focus on creating a bond between the recruit and the group by instilling a sense of belonging and in extension loyalty. This became the basis for a second element of initiation that focused on motivating the recruits as fighters and imbuing them with fearlessness in the face of battle.

BONDING THROUGH TRIAL

The bond forged between initiates through their shared experience, came out of their ability to face and overcome a series of trials that made up initiation. A number of levels of initiation offered increasing powers, but at an equally increasing cost.

At a basic level, initiates were brought into the group, allowing them to effectively substitute their family—that in many cases was being left behind to go into battle—with the society:

> They use native herbs, which you drink and they bathe you in. These herbs make it so you don't feel like returning to your family.[39]

This was particularly important for the younger recruits, some of whom would not have been through the Poro initiation despite this desired standard, either because of their age, lack of means, or because it had been disrupted by the war. It reflects the bonds forged through the Poro and Bondo societies, although as with so many things at times of war, a more pronounced version was adopted.

Whereas in the RUF part of the initiation of particularly forcibly recruited fighters often involved the public killing of a fellow recruit or civilian, the Kamajors used their own blood rituals to strengthen ties between fighters. This generally involved ceremonial cuts and markings, and corporeal beating. A detailed account of the initiation process at a basic level was given by a former child-combatant to the Special Court for Sierra Leone. According to his account his initiation took place at Base Zero together with 400 other recruits:

> When we went into the Society bush, the others who were there for us to enter the bush, we entered the bush. We were stripped naked, all of us. We only had on our pants. Then we started singing. After the song, they brought a razor blade.
>
> When we entered the bush and we stripped naked, there was a razor blade and they started putting some marks on our bodies. After they had marked—placed these marks on our bodies, there was something in a drum, which was black.
>
> They gave me that thing that was in that drum—that black thing.
>
> The thing that was in a drum, we would go there and take it and smear it on our bodies.
>
> When we had smeared it on our bodies, they told us not to bathe for one week.
>
> After the one week had elapsed, at night, at about 2.00 o'clock in the night they took us to a graveyard.

> We were there until 4.00 o'clock. Then we came back into the bush.
>
> We were told that if anybody sleeps there you would be sleeping—if anybody had died for you, whether it's your grandfather or any of your family members, he would come and give you something, which would make you very powerful to fight. That's why they took us to the graveyard.
>
> ...we came back into the bush. Then we had our bath. Then it was in the morning.
>
> ...on the morning after we had all had our bath, we came back into the bush, then there was something else which we had to do. 40 people were standing on one side and another 40 on the other side. All of them had different canes in their hands.
>
> We ran in between those two lines. As we were running, we were being beaten.
>
> They had different canes. They would come from up—running in a line, and they would start whipping us. As they are beating us, some people fell and collapsed. As for me, I had some swellings on my body.
>
> After they had beaten us, they made something inside a drum, which they called in Mende nesi—it's a potion.
>
> We all brought a rubber—a sass-man rubber, and it was put in it, but before they could put it in the rubber, they told us that that potion is our protection and, when we go to fight, that's what we would smear on our body for the fight, together with a ronko. That's what was given to us in the Society bush.[40]

The level of initiation determined the types and extent of rituals used as well as the powers acquired. Some initiates reported how they declined higher level initiation to avoid more advanced rituals said to involve the risk of "madness and insanity."[41] The severity of the shared experience that constituted initiation was intended to forge an ever stronger bond, at the peak of which there was no turning back.

The idea of "no turning back" is a theme that runs throughout initiation. Referred to as *baa woteh* and translating as "do not turn" it reflects the commitment to not only face battle bravely and without succumbing to the temptation to run away, but also a commitment to the group and community which precludes any disrespect or abuse of civilians and requires absolute loyalty.[42] Instilling that sense of loyalty and commitment was the main task of initiation, in addition to the distribution of immunity.

Physical harm was an integral part of this process. As part of initiation, initiates would typically be marked with cuts from a razor, and go through a trial involving beatings. Some accounts tell of recruits falling victim to rituals that made up part of the initiation of the rest, or local people disappearing during the times of initiation. Pregnant women were said to be particularly at risk because of the potency of body parts associated with the unborn child such as the umbilical cord. The sign of recognition—or the shared secret that could be used for identification—between Kamajors was said by some to be a dried umbilical cord tied around the head of the fighter for protection as part of the initiation ceremony.[43]

Although verification of these rituals is difficult, it illustrates the power associated with the initiation process—both through the severity of the actions required to complete it, and also the spiritual power supposedly acquired through the rituals. In the eyes of the local population, the disappearance of a pregnant woman (whether or not she had in fact been kidnapped or mutilated) could equal spiritual power gained by a society member, and in extension sustain a belief in the power and strength of the Kamajors. The suggestive power alone must not be underestimated, especially its ability to feed the sense of invincibility of the fighters who operate in a context where these beliefs are widely held. Cannibalistic practices in and around battle have, however, been more widely reported and verified and lend greater credence also to the use of such practices during initiation.[44]

The higher level initiates received both more potent and lasting powers. "They have something in their stomachs," and the power that emanates from this substance stays with them for life, even if it is no longer put to use in conflict and war.[45] Again, the traditional image of ingestion is referenced to denote the spiritual powers of the higher initiates; just like the Burkinabe RUF fighters were thought to possess such substances, powerful Kamajors were put on a par with their own, lasting physical—if unseen—source of power. Recruits who received protection through basic initiation and the administering of potions and charms, however, only got temporary protection that waned with the medicine that provided it. The higher levels of initiation and their associated powers were seen not only as a privilege but also in part as a sacrifice as the powers came at a price—the acquiring of knowledge and skills that could drive a man insane as much as provide powers. It was because there was no going back from these levels that some chose deliberately not to progress beyond the more basic initiation.[46]

The levels of initiation and resulting differentiation of powers were further entrenched as recruitment became more liberal and the selection of initiates less regulated by traditional authorities in the local communities. New members were differentiated from "old Kamajors" with the suggestion that the original and early initiates were more pure and as a result had greater power.[47]

Rewarding Loyalty with Magic

The second element to the initiation "rewarded" the recruit for his loyalty to the group by awarding him magical protection and power. Immunity to enemy fire or other means of bodily harm is the most cited, and probably the most "effective." This "medicine" was introduced in the myths of the Kamajor society's origin, and is a variation of the medicine practiced by morimen and medicine men among the Mende in general. Becoming immune to bullets and knives on its own allows for greater daring on the battle field, but it is supplanted with rituals specifically intended to instill fearlessness in the fighters:

> The initiator placed charms around my neck so that bullets are deflected, and a special charm to protect against knives and pangas [machetes]; they will just bounce off my body. After this they gave me my special dress and herbs and charms. I was instructed not to bathe for several days because it would reduce the magic. Later, they rubbed the blood of a human being on my skin, and I found that I was not afraid of anything. I had a strong and fearless heart.[48]

Accounts of bullets bouncing off Kamajors' bodies in battle are numerous. One ex-combatant now in parliament explained how he could feel the bullets hit his body and just fall to the ground.[49] Often specially prepared *nesi* would be distributed to fighters before battle to "top up" protection. In combination with the use of medicines made in part from cocaine derived or other natural narcotics, the rituals were given immediate effect. Other accounts, as seen above, describe a more psychological approach to the same desired effect where initiates would be asked to lay down in a cemetery at night to remove their fear—if they got up or ran out of fear they would be killed.[50]

At the end of the initiation process, the powers gained by recruits would be demonstrated in a final test that saw live rounds of ammunition fired at the fighters.[51] Although much testimony has been given to the verity of the rounds being live, evidence from Liberia where blanks were fired at NPFL recruits suggests alternative explanations. However, the fact that some recruits were reportedly hit would imply that there was live ammunition involved. Incompetent use of weapons and the poor maintenance of the same could allow live rounds to be fired with relatively limited damage.[52] In terms of the function of this final test, it proved to recruits that their initiation had worked and imbued them with the power of the Kamajor, but also weeded out the "infiltrators and people associated with witchcraft"—ostensibly the people who were hit by the bullets.[53] This restriction of power to those who were "pure" comes again in battle where powers were thought to fail as a result of violations of behavioral rules given to fighters.

RITUAL AND GROUP COHESION

It is clear that the practice of initiation played a crucial role in the development of the Kamajor society into a big player in the civil war. It offered people security and support and a means of overcoming fears of death in combat. From the perspective of the society and armed group it helped bond fighters together and instill in them both a sense of loyalty and in extension discipline.

As noted in chapter 6, Paul Richards has drawn on Durkheim's ideas about religious life to situate ritual in the Sierra Leone war as a means of creating group cohesion. This perspective rests on a view of religion that sees its center of gravity in practices rather than beliefs. Whereas Richard's main object of study is the RUF where practice may indeed have quickly overtaken ideas and beliefs in importance, relegating ideas and beliefs to a secondary role does not hold up easily in the case of the Kamajors. As argued earlier in relation to the creation of practice, the cosmology of the agents involved in the constitution of practice cannot be seen as separate to the functions it serves, but rather forms a whole that must be considered in its entirety for the practice to make sense. In the case of initiation, belief is a crucial part to the experience and the function it serves. But the social functions of the rituals that emanate from these beliefs certainly reinforce the mobilizational capacity of the "system" (including in this case belief *and* practice), which is why Richards' discussion is relevant also to the case of the Kamajors.

Richards compares the sodalities of Roman times and Medieval Europe with the fighting groups in Sierra Leone, and whereas his main emphasis is on the RUF, this comparison is perhaps more poignant with the CDF:

> Roman and medieval sodalities were associations for craft and trade, but never simply functional. There was always a religious dimension. Members were admitted through initiation, and groups sacrificed regularly to tutelary deities. Craft, and the craft of ritual, went hand-in-hand.[54]

Richards' conclusion is that "the sodality is a recurrent convenience of small-group sociology" and that as a result "rite creates group."[55]

The creation of a group through initiation and continued ritual practice certainly applies to the CDF. And although the ritual practice is crucially underpinned by the belief in the potency of the powers administered, the process of illuminating those powers and the access to them through ritual is what allows for the creation of the particular group. The beliefs and the rituals that embody them together generate solidarity by enabling collective action, but also serve wider existential functions by answering questions about meaning and causality. The role of ritual must be seen as contributing to this wider context and shaping of behavior and action, and as a tool for directing activity in an effective manner.

One senior Kamajor commander acknowledged that the bonding experience of initiation was crucial; you needed to have been part of this bonding experience to appreciate how the members thought about and would react to things.[56] When you did, as a commander you could wield great power and control over your men. Although this might sound like a psychological profiling and shaping exercise that allows a commander to manipulate his men once he has molded their perceptions of the world, it is rather a recognition of the seamless blend of magical belief, ritual practice, and practical managerial and disciplinary considerations. In this case, it was made clear from discussions that the commander shared the belief of his men rather than attempted to encourage them for his own functional benefit. In other words, these fields are not seen as separate and distinct, they are all parts to the shared and accepted reality (or in Bourdieu's words, habitus) in which action takes place.

The traditional appeal of the Kamajors was closely tied to the privileged place they held in Mende society. During the war this was supplemented by an acute need to protect one's family and community as well as a means of survival. It was further strengthened by the emergence of myths of invincibility that made particular reference to the ongoing conflict and the enemy, building on already established perceptions of spiritual might. In this way, the society both fit into the existing habitus and molded it to respond to the evolving field of conflict.

The process of initiation grew out of this heritage and allowed the civil defense militia to develop into a popular and cohesive movement. Initiation provided fighters with a sense of shared purpose through the bonding that took place while undergoing the ritual process. In a manner similar to that in which the Poro and Bondo societies shepherded youth into adulthood and created networks that would remain powerful for life, Kamajor initiation turned recruits into fighters with similarly strong bonds and loyalties. The exercise was in part a function simply of the severity of the situation and the shared experience; particularly where recruits faced physical pain and fear. But the psychological bonding exercise was complemented and reinforced by the spiritual powers and magic employed to equip them as fighters and bind them to their duties. The functions served by initiation were readily acknowledged by the leaders as both providing the physical protection promised—through spiritual power—and motivating fighters to go into battle. Hinga Norman said in his testimony to the Special Court that

> Initiation means going through the rites of training and this is performed even in the conventional western warfare training...In our own tradition,

> you are segregated for a period of either days, weeks or months and certain thinks are taught and certain situations develop for you to go through, so that you cannot turn tail, run away in the face of battle. What is then given, particularly to immunize you, to make you even a lot bolder, is called immunisation.[57]

In this sense, the practice is not dissimilar from any other bonding exercise undertaken as part of military training to build a cohesive fighting unit. But because of the particular worldview shared in by the individuals partaking in this exercise, it is combined with yet another layer of practice—the immunization. Norman acknowledges the function and its similarities to that achieved through other means elsewhere, but not without maintaining the actual power of the magical element involved in it: "The simplest form of immunization in the western world . . . is the bullet-proof garment. In the olden days the warriors wear iron shield [*sic*]. The iron shield of the initiated kamajor is a means by which nothing is worn, but one is safe from missile from head to sole [*sic*]."[58]

The two dimensions to initiation—the shared experience that strengthened ties within the group and the motivation this generated together with the magical protection imbued—tied together by the figure of the initiator presiding over both, formed a core of the Kamajor society that attracted recruits and both admiration and fear from noninitiates. It set the fighters up for duty whether in action on the battlefield, or as intelligence gatherers or reserves in local communities.

In a context where belief in a spiritual reality and associated powers is widely shared it was only natural that this reality became incorporated in the activities undertaken by the people sharing in it. Under these circumstances, magic served a number of functions, in part through its use and in part through belief in its effectiveness. It specifically aided mobilization and combat motivation and, to an extent, the maintenance of internal discipline. For the Kamajors, mobilization was structured around the organizing principle of the secret societies already firmly established as part of the local communities where they sought recruits. Whereas in peacetime, adolescent boys and girls would go through initiation in the Poro and Bondo societies respectively to enter adulthood, during the war a different society was set up to transform civilian men into warriors. The purpose was different, as was the process of initiation, but the structure was familiar. Much of the set-up was equally familiar to the Poro and Bondo initiation; initiation itself required seclusion and the spiritual specialists presiding over the process were from the same professional corps as those looking after matters of initiation in peacetime. The main difference was the immediate purpose and the powers bestowed on the initiates. War required greater power and protection. The initiators rose to this challenge and devised several new rituals particularly tailored to the task at hand. But like a blacksmith turning from the mundane duties of shoeing horses and sharpening scythes to manufacturing weapons of war, the initiators directed skills they already possessed to put together the package for wartime immunization.

In other parts of Africa we have seen a similar adaptation of practices already entrenched in a society for wartime purpose. It is the natural utilization of available resources to meet a new challenge. In the same way, spirit mediums and witchdoctors have been called upon to provide services in other times of turmoil, as documented by Geschiere and Masquelier amongst others with reference to political and economic pressures.[59] But although it is therefore tempting to jump to a conclusion that reflects a structuralist approach to both religion and human behavior, the relationship between context (challenge) and practice (chosen action to face that challenge) is more complex than that. Cosmology no doubt becomes increasingly important in times of adversity as both explanation for misfortune and avenue for redress, but the relationship between agent and action is deeper than the simple exchange of service. Similar to an ailing customer entering a pharmacy it involves the impetus for the sense that there *is* a remedy as well as the knowledge of where to look for it. This impetus does not emerge at the time of crisis but out of a wider conceptual space occupied by the agent before the challenge emerged. In the case of the Kamajors, by the time the crisis arrived, the spiritual realm was already a natural point of reference.

CHAPTER 9

The Kamajors in Battle: Magic, Tactics, and Success

> You have to know the society to make the society members perform.[1]
>
> A group of Kamajor fighters were to be deployed by helicopter, and a pilot arrived to pick them up. The Kamajors had gone through their pre-battle ritual to ensure their protection was absolute. Part of this ritual had involved drinking a potion especially prepared by one of the initiators. The fighters were all very agitated and eager to get on the helicopter and onto their destination. Eighty Kamajors got on the helicopter that would normally seat twenty-seven. Significantly overloaded, the helicopter was too heavy to take off with all the fighters onboard. But no Kamajor wanted to be left behind or have to wait for the helicopter to return, and all refused to leave the helicopter. Not even their commanders could convince them to stand down in their agitated state. The helicopter pilot had to wait for two hours before the fighters had calmed down enough to be persuaded to leave.[2]
>
> A group of Kamajors were in battle together with some ECOMOG troops. One Kamajor was in a van that held a gallon tank of magic water—a potion for protection in battle. The van also held several similar tanks full of diesel. As they came under fire, ECOMOG soldiers came running towards the van crying "give me the magic water". They had seen how the van was left unharmed by the bullets, like a reversed magnet repelling the fire, and they too wanted to get the same protection. They came into the van, but confused the tanks and took ones that contained diesel, not the magic water. But they thought they were protected and went back into battle stronger and fiercer. The Kamajor stayed in the van with the real protection.[3]

These accounts of events that took place in combat situations paint both complementary and contradictory pictures of the role played by ritual and magic. On the one hand, it suggests that the use of drug infused potions altered the state of mind of the fighters and made them fiercer and braver, and in the case described above, more difficult to command while under the influence of the potion. On the other hand it suggests a very strong psychological dimension, something of a placebo effect that allowed the ECOMOG soldiers to use the diesel instead of the magic water and, feeling sure of its power, face battle as fearlessly as had they had actual magical protection. The second story was, interestingly, told by the Kamajor in the van, who wanted to demonstrate the power of belief amongst non-Kamajors as well as within the Kamajor ranks. But although he recognized the tangible psychological effect the diesel had on the ECOMOG soldiers, he also maintained that the real protection stayed with him in the van where the magic water was kept.

This combination of a powerful belief system, coupled with the use of medicines and potions, formed the basis of Kamajor battlefield magic. Its working and success was highly dependent on the initiation experience and the bonds that were forged and beliefs reinforced through it. Magic specifically designed for combat grew out of this and was a logical extension of the wider cosmology of the fighters and their opponents as well as the knowledge imparted to initiates during initiation. The mandate of the initiators and the commanders became a powerful one as they could bestow on their men the power to do things they would perhaps have been unlikely to do under different circumstances. Hinging on the maintenance of purity, loyalty, and discipline, the magic created a strong narrative of justice as well as power.

The Powers and the Taboos—the Enabling and Limiting Habitus

Immunity and the rules that were required to be upheld to ensure its effectiveness, familiar from the initiation experience, reappear on the battlefield. Together with ritualistic practice that further aided the deflection of enemy attack and informed strategic and tactical decisions, this formed the magical dimension of Kamajor fighting.

Before going into battle, fighters were reminded of their protection and what was expected of them as a result, strengthening their resolve and providing a powerful mood enhancer in the face of a life threatening situation. Gathering fighters before an operation at Base Zero, the High Priest urged them not to be afraid and to remember their powers and protection:

> Mr Allieu Kondewa told us that all these powers that he has in him has been transferred to us so that nothing will be wrong with us, no cutlass will strike us. He's now satisfied. So all of us will go to the war front and come back with happiness, and let no one be afraid.[4]

Numerous powers are ascribed to the Kamajors. The most widely known is their alleged ability to become immune to bullets as a result of their initiation and subsequently through protective charms and clothing worn in battle. But depending on the level of initiation, their powers extended far beyond mere protection and became increasingly proactive: Fighters were said to have the power to become invisible and to shape-shift. They could grow as tall as the trees to see the enemy in the distance, and could penetrate walls and other solid structures without the need for doors. The latter was recounted by civilians who explained that at times this was how the leaders and initiators claimed the victims they needed for certain rituals.[5] On the battlefield, their powers not only allowed them to track the enemy and attack under the cover of invisibility, they could also direct the paths of bullets and other objects. "Controllers" would lead a unit's way into battle and deflect enemy fire. "Cut offs" would completely silence the enemy weapons if the fighting got too intense. One Kamajor ex-combatant explained their roles in a radio interview with Radio Netherlands:

> We have something within the *Kamajor* we call "control" and "cut-off." Whenever we went into battle, the person who was the controller would go in front. He would deflect the enemy's heavy artillery, throw it over us and let it fall somewhere else. The other one, the cut-off, we used him when the tension was very high. He would make sure all their guns were silenced. No rebel would be able to fire his weapon when the cut-off did this.[6]

At times, the controllers would be positioned on the roofs of houses and direct the battle from there if a village was under attack. In a similar way, any object could be controlled by a powerful Kamajor. One famous incident involved a plane used by the junta to transport weapons that were allegedly brought down by a Kamajor who simply shouted at it to fall. The problem with the plane was said to be impossible to determine let alone repair.[7] Enemy fire could also be controlled and directed by specific tools such as the "RPG-7 bat." This "bat" consisted of a piece of mirrored glass that would reflect and subsequently deflect any incoming RPG fire.[8]

Another commander described how he put an entire camp of RUF fighters to sleep before attacking them, to allow his relatively small force to defeat the more numerous rebels.[9] Similar powers of mind-control were used to infiltrate enemy camps for reconnaissance missions.

All these powers, and their successful application, were intimately tied to the behavior of the fighters. The adherence to the code of conduct is what allows the magic to work. *Baa woteh*—the principle of no return—remained central to the strength of the protection but was supplemented with a range of additional taboos: Typically fighters were not allowed intercourse while on active duty, could only eat certain types of food, always had to move in the direction of the fire, and could not use any type of drugs.[10] One commander explained how before battle he would sleep with his fighters in a church or a mosque where no women could enter. Instead the women would bring food and leave it by the door. The combination of the holy place and the seclusion from women helped keep the fighters pure. On the day of the battle the fighters would have to wash carefully before engaging the enemy. This also led to a practice of keeping buckets of water at hand or staying close to rivers to allow fighters to wash quickly if surprised by the enemy.[11] Although the rules differed somewhat from account to account, the essence of them remained the same—the maintenance of a pure body and mind.

Failure to abide by these rules would nullify the immunization and the fighters would be mortally wounded by bullets. It is this caveat that allowed for the survival of the beliefs and practices in spite of the deaths of many initiated fighters on the battlefield. Individual soldiers explain their successes in battle both in terms of the magical protection bestowed upon them and their ability to abide by the rules:

> I was never injured in the fighting. I was really strong because of our medicine. I never "spoiled" (violated) the laws, so I never got injured. But many of my friends were injured because they spoiled the *Kamajor* laws.[12]

At the opposite end, as indicated by this fighter, those who were injured or killed were assumed to have broken the laws. The breakdown in discipline and authority structures as the war wore on and food and resources became increasingly scarce was also explained in these terms: Many fighters took to drugs to counter the horrors of the fighting, but also the hunger and cold, and in the process breaking one of the laws that ensured ritual protection. Similarly, the liberalization of initiation during the mass-recruitment stages of the war, and the inclusion on noninitiated fighters in the ranks, compromised the purity of the Kamajors and jeopardized the powers they had acquired. It was as a result of these transgressions that they explained the increase in Kamajor deaths.

One ex-combatant explained how the movement became divided between "old" Kamajors who had been initiated in the early stages through the original process, and "new" Kamajors who had been through brief initiation and viewed the process as a commercially acquired commodity not dependent on any particular type of behavior. This attitude was fuelled by the services offered by some initiators (including the High Priest Kondewa) where for the right price fighters would be initiated to a high level *and* be absolved of the taboos normally associated with the powers they acquired.[13]

But as much as the "corruption" of practice tainted the movement and particularly its reputation and image in the eyes of the civilian population, it did not undermine belief in the real power involved. Similar to examples of spiritual practice elsewhere, as seen in chapter 3, these powers could be constructive or destructive, and although Kamajor practice in many instances turned destructive, it did not follow that people stopped believing in their power per se. It did, however, present a problem to the Kamajors and as a result a series of restructurings and changes took place to attempt to return the "pure" practice and regain the trust of the civilians and traditional authorities.[14]

CANNIBALISM AND INTIMIDATION THROUGH ECHOES OF TRADITION

There were also ritualistic elements to the fighting, and behavior engaged in by the Kamajors during or after battle, that appear to have been more of an echo of organized ritual than actual orthodox practice. Acts of cannibalism and ritualistic killings are the most striking of these "echoes." It is important to distinguish between organized ritual and distorted reinterpretation of them while at the same time recognizing the shared context that they grew out of.

The most disputed element of the Kamajors' battlefield behavior was the cannibalistic practices some were accused of engaging in. More so than any of the other ritual or ritual-related dimensions to the Kamajor movement, cannibalism, and to a lesser extent ritual killings, are often disputed by ex-combatants. Nevertheless, their occurrence has been documented reliably in many instances.[15] Kamajor fighters on several occasions dismembered enemy victims and ingested parts of their bodies—particularly the heart and liver. The symbolism of the practice—or the tradition echoed—is the acquisition of power through ingestion. In the case of a warrior, the power of the enemy is acquired by ingesting his heart or liver. This resonates with more broadly held beliefs, in West Africa and beyond, in the physical body as a repository for spiritual power. But the manner in which this was practiced by Kamajors during the civil war was a reflection at most of the more traditional rituals that have featured in initiation societies in the region. The transgression in particular of the element of secrecy that permeates initiation society practices indicates both ignorant improvisation and a primary desire for intimidation and the immediate benefits that can be drawn from it.

One Kamajor group notorious for their cannibalistic practices was the Yamorto squad. This group was allegedly characterized in part by their habit of cannibalizing enemy victims. One Kamajor ex-combatant described to the Special Court how they would eat enemies once executed:

> We had another group, which was called the Yamorto.
>
> Yamorto, it was a group that was meant to eat people.
>
> When we entered Kenema, when we caught the collaborators, we tie them with an FM [a stick and rope contraption used to tie the hands of captives], and we would ask them—say, "You were with the rebels," and they were doing business with them; and they would deny that, and they would say they were just with them in the town. As time went on, after we had tied them with that FM, as we tie them, we'd tell them to tell us the truth. And they would tell us that they were with the rebels, but they weren't carrying guns. After they had told us that, we would take them to the base, which was the Yamorto base.
>
> When we took them there, in most cases it was to eat them at the base.
>
> Yamorto squad were there to eat any collaborator. That was what we were.
>
> When we captured them, we tied them to a stick with a tong. From there here was a commander called Colonel Biko. He would come. He would give us the order to put the man on the stick with the tong. Where our base was was close to a swamp. There we take the person to be killed. When we take him there, when we reached, we hold him, we put him on the ground. From there we start to choke him with a bayonet, then he will die. When he die, then the heart, the liver and the other parts in his stomach we remove it and the legs. Then the head, we find a stick and put it on it. We take it to the gate. After that we would come. We heat some water and remove the body parts in bits and place it the hot water, then remove the first skin. After that we have oil. We place that over the fire. After that those body parts which had been removed and the skin removed as well, we place it in the oil and fry it. After that we prepared some gravy. Some people would eat it with bread.[16]

This account displays a striking blend of the horrific and the everyday. The act of cannibalism and its ritual significance is "normalized" through a strikingly mundane cooking process. It is certainly possible that this added to the intimidation by treating the victims like any other meal, but it is also possible that the cooking allowed those eating to dissociate their meal at least in part from the substance of the main ingredients.

But cannibalistic acts also took place on the battlefield without the seclusion of a camp or the benefit of condiments. As seen in Liberia, Kamajors from Sierra Leone engaged in the killing and ingestion of enemies and assumed collaborators in plain view and immediately upon capture.[17] This is more reminiscent of the acts of violence and atrocities committed by the

RUF where the public spectacle of the act in itself provided part of the gratification for the fighters seeking power—be it through subjugation of people through fear or through the actual acquisition of magical power.

THE NATURE OF FIGHTING—WEAPONS AND APPEARANCE AS TACTICS

When looking at the way magic and ritual was used in combat it is important to take into account the general nature of battle and fighting during the war. In the interaction of fighters and forces we see the creation of patterns and contexts that both feed into and feed off each other. The practices combine to form a particular battle-space in which magic had a role to play. The context of the war and battle is shared but also constituted by both fighting parties, allowing their actions to be mutually intelligible and as a result both meaningful and effective.

The style of combat in Sierra Leone is one that is familiar from across the continent. Alternately described as "inefficient," "disorganized" or with a suggestion of cowardliness and laziness, some have observed more accurately that the particular form of fighting is the result of some very clear constraints imposed by the environment—human and geographical. Battles as we have come to think of them in the West are virtually nonexistent. That is to say, pitched battles between clearly defined forces engaging each other directly and with lethal intent. Instead, warfare in Sierra Leone tended to revolve around ambushes, tended to aim not so much at the physical destruction of the enemy but his retreating and ideally dispersing. The main aim is to induce the enemy force to relinquish their position and withdraw, leaving behind as many of their weapons as possible.[18] As a result, although killing is by no means uncommon, the main aim of firing weapons is not to kill as many of the enemy as possible but rather to display the superior power of one's own force and induce the enemy to flee—some have rather expressively referred to this aim as "making as much noise as possible."[19] This also explains the predilection for weapons such as the AK-47, RPG-7s and mortars that are particularly suitable for making noise and displaying firepower, but less useful when engaging fighters in the forest if the aim is to actually hit them. This is particularly the case if little or no training on how to use the weapons has been given.[20]

The example of the Kamajor attack on Koribundu in February 1998 is an illustration of this. The Kamajors attacked Koribundu on Friday 13 February. Attacking from three directions more or less simultaneously (see chapter 5 for a discussion of some of the planning of this attack), the Kamajors nonetheless faced heavy resistance from the AFRC and RUF in Koribundu and retreated after only about ten minutes of fighting. In his expert witness statement to the Special Court, Col. Richard Iron concluded that "they appear to have given up at the first sign of resistance, perhaps within five minutes of coming under fire."[21] But in spite of this absence of a real fight, the AFRC/RUF retreated from Koribundu during the night of the 13th leaving the town free for the Kamajors on the 14th. The reason for the AFRC/RUF retreat remains somewhat unclear, but Iron suggests that they received news about the fall of Freetown to ECOMOG and decided to retreat to regroup and join their other forces further north.[22] But because of the general lack of pitched battles, this retreat could still be claimed by the Kamajors as a success, rather than as a humiliation or defeat—the enemy had withdrawn, and the Kamajors came to be in possession of the town they had set out to take, even though they had withdrawn from battle.

Interestingly, the Kamajors had reportedly prepared for the attack on Koribundu in part by planting a magic talisman in the town during a reconnaissance mission.[23] The talisman was intended to cause friction within the AFRC/RUF camp and make it easier for the Kamajors to successfully drive them from the town. Reports that trouble was indeed brewing within the AFRC/RUF camp signaled to the Kamajors that the time was ripe for attack. It is therefore possible that the eventual retreat of the AFRC/RUF was in part interpreted as a result of the nonmilitary weapon used by the Kamajors to undermine the enemy's strength from within, further helping to see the unfolding of events not as a defeat but as a victory. This was certainly the perception of the Kamajors who saw their capture of Koribundu as a success.[24]

The importance of appearance is further reinforced by the tendency of fighters to dress in elaborate combat garb. Instead of uniforms or fatigues that serve to camouflage fighters, the tendency is toward emblems of power. For the RUF these tended to be a mixture of accessories associated with popular culture personae (Rambo is the favorite cited by observers, but hip-hop and gangsta rap personalities also featured in their repertoire) and the occasional pastiche of traditional symbols (the masks of secret society traditions were one of the props reinterpreted by some fighters to add to their war persona). For the Kamajors the points of reference were more exclusively traditional, and fighters were generally covered in amulets, charms, and specially prepared protective clothing. For the Black December operation, many Kamajors wore black uniforms to add to the fierce appearance of their forces as they entered towns.

Both the sound and sight of the fighters were central to their style of fighting. Why? Some have argued that in a region where communities are relatively small but closely interlinked, the idea of pitched battles and heavy casualties would amount to fratricide.[25] From this perspective, attempts to subjugate the enemy by less than lethal means would be preferable to ensure the postconflict ability to rebuild society and return to normal life. There is certainly some merit to this argument, but it is equally important to not overlook or dismiss the brutal violence inflicted on the country by the war and the acts of the combatants. Crucially, however, much of this violence was inflicted not on the enemy forces but on the civilian population. In part directly, as punishment for perceived collaboration or purely for purposes of intimidation, and in part indirectly as much of the weapons fire that did not hit the enemy hit civilians instead. This was particularly the case with mortar fire but also with straying AK-47 and lightweight machinegun fire.

"Compelling the enemy to do your will" therefore takes on a somewhat different guise in the Sierra Leone context than what would be expected from a more traditional Clausewitzean perspective.

In many ways, the war in Sierra Leone was fought with guerrilla tactics by all sides. Some corrupted and disfigured traditional guerrilla approaches such as winning local civilian support (notably the RUF and the AFRC). Hoffman has noted that "One of the other logics at work here is the logic of power in this region of Africa. As the Sierra Leonean commander explained it, a 'hearts and minds' guerrilla campaign, aimed at mobilizing popular support, had little purchase in a West African context."[26] Quoting a Kamajor commander, he illustrates the perception of the rebels' approach to guerrilla tactics:

> [Targeting civilians] is one of the major tools in guerrilla warfare. Because when the guerrilla is fighting, he is less equipped, he has less manpower. He's going to use tactics to put fear into the civilian populace and send the signal to the government that it can't protect its people... It is one of the tools the guerrilla uses. Fear and intimidation.[27]

But also military tactics revolving around ambushes and hit and run attacks, instilling fear and a sense of helplessness in the enemy and slowly eroding their will to resist, were common throughout the war. The Kamajors in particular made use of their knowledge of local terrain to stage ambushes and surprise attacks. Although attacks against enemy positions were rarely successful unless launched with significantly superior numbers, as seen in the example of Koribundu given above, ambushes drew on the Kamajor strengths and tended to be successes—successes in the sense that they usually intimidated the enemy and as a result disrupted their transport and communications routes by diverting them away from a given area or road, and in the sense that enemy weapons were often captured in this way.

Ambushes succeed because of three combined factors; superior knowledge of the terrain, good intelligence and the successful execution of surprise. The hunting tradition, but more importantly the knowledge of the forest and region, helped the Kamajors achieve all three. Good relations with the local communities helped intelligence gathering, as did the significant skill of many Kamajors at infiltrating enemy strongholds to gather information. Crucially, all three were associated both in the minds of the Kamajors and the wider imagination of the civilians and enemy with the secret knowledge and power acquired through the Kamajor society and its initiation.

Sharing the Battle Space—Shared Belief and the Space for Magic

The Kamajors and the RUF were both military forces and at the most basic level they engaged each other militarily. Although, as often as not, the two forces battled for space and control without actually engaging each other directly. The emergence of the civil defense militias and particularly the Kamajors severely threatened the continued survival of the RUF and posed the first serious threat to their campaign. One outcome of the initial tactical superiority of the Kamajors was therefore the RUF's increased brutality toward civilians to "break the cooperative link between rural civilians and the civil defense militia."[28] This began a circle of increased violence as the augmented levels of atrocity by the RUF provoked greater support for the Kamajors and in turn a greater need to undermine their support-base through additional violence. But the interaction between Kamajors and the RUF played out on additional levels as well.

The usefulness for Kamajors of incorporating magic and ritual in combat behavior was in part derived from the shared perception of spiritual power of the Kamajors and their enemies. Their rituals were not isolated practices purely for the benefit of Kamajor recruits, but resonated beyond the own group. And as such they gained an additional level of potency. Therefore, to understand the impact of Kamajor use of magic and ritual we have to recognize the role and perception of the same amongst their enemies.

As the RUF drew its recruits (voluntary and coerced) from across the country they embodied a wider range of religious beliefs than the Mende-centered Kamajors.[29] Their organization could not be pinned to a preexisting framework or tradition as the Kamajors' was to the hunting society (albeit in name more than actual practice). The RUF also saw itself as the antithesis of tradition, ushering in a new order, different from what had gone before them. Nonetheless they had to accommodate the beliefs of its fighters—leadership and foot soldiers—and originally incorporated in its regime the beliefs and practices common to Sierra Leoneans. However, although herbalist protection and supernatural powers were sought by the RUF in the first years of the war, they subsequently appear to have been consciously dropped by the leadership as central practice. Peters suggests that after initially being part of the pre-battle preparation, the RUF decided to abandon the use of magic as the protection offered by *kamohs* proved ineffective. He quotes one RUF commander saying;

> You know, the RUF already used magic in 1991. But then we decided to abandon it the same year. You know why? Because we had too many casualties in the frontline. So we called upon these "kamo's", who so-called initiated us, to get this bulletproof, to put their best protection on themselves. After that we shot [at] them. All but two of them died. So we abandoned it. We are mathematicians, we need to have proof.[30]

According to another RUF commander, Sankoh condemned all superstition in 1995 and taught rebels to use tactics designed to intimidate people and play on their spiritual beliefs.[31] But while there was clearly expressed skepticism on the part of some commanders, it is difficult to say to what extent this took root amongst the rank and file who may not have believed in their own immortality but still showed fear of the Kamajors.

Despite the sidelining of initiators, religious practice retained an important role in RUF camps, and the lack of religious observance could result in severe punishment. A female ex-combatant described the central role of religion in RUF camp life, pointing to frequent and regular religious worship for both Muslims and Christians in the camp: Muslims would keep the fast during Ramadan and Christians would pray in church. She also reports, in some contradiction to the testimony mentioned above, how a moriman from Guinea helped the fighters prepare for battle in a way similar to their function in the Kamajor CDF.[32] Although the religious rituals were not an organizing structure in the way it was with the Kamajors, the RUF drew on the same beliefs and practices as they resonated profoundly with its fighters.

The main contrast with the Kamajors is that the RUF did not use religious ritual as an organizing principle or structure for recruitment. This reflects the heterogeneous nature of the RUF and its lack of any one community base. Its dissociation from traditional authority and community configurations left it dependent on outside or independent spiritual guidance.[33] It also placed greater emphasis on the individual powers of its leader Foday Sankoh than did the Kamajors who had several powerful spiritual leaders, some, but not all, of which were also military leaders. Sankoh's powers were perceived as far-reaching both by his own fighters, civilians, and his opponents, and were reinforced by the belief that he gathered particular spiritual strength while abroad in Ghana, Libya and Liberia. He had in his service for some time one of Charles Taylor's medicine men, Alhaji Kuyateh, and could allegedly communicate with his forces over vast expanses and summon reinforcement from Liberia by supernatural means. In a similar manner to the Kamajors, the RUF's myth of origin also involved spiritual reference, as Sankoh was said to have been commissioned by God to bring down the corrupt regime in Freetown.[34]

Religion and ritual became more important for fighters and the group after recruitment into the RUF and in the everyday life in the camps and in battle. This does not alter the strength of the beliefs but reflects the context of the fighting force and the recruitment strategy this resulted in. It is perhaps also the more common path for groups distancing themselves from local communities than those anchored in them, and organized as *defense* forces as opposed to *revolutionary* forces challenging the local and national political reality. While it can be more difficult for a revolutionary force—particularly one like the RUF that tends to target as opposed to support civilian populations—to attract recruits, once new fighters have been incorporated into the ranks they must be conditioned into effective ones. Tapping into the recruits' worldview and, as part of this, religious beliefs, this can be done with greater success. A similar approach has been taken by the Lord's Resistance Army in Uganda, where abduction has been the main means of "recruitment" over the past decade of their struggle against the Ugandan government. However, once abducted, the new fighters are put through initiation rituals and given a behavioral code heavily referenced to traditions anchored to the spiritual world.[35]

The shared cultural context and sets of meanings (habitus) meant not only that the initiation into the Kamajor society resonated with noninitiates; it also fuelled the construction of initiation practices in the RUF. Although RUF initiation appears to have been less focused on spiritual power and protection, they served similar social functions by substituting the traditional secret society initiation (and consequent graduation to adulthood and associated community responsibilities) with a set of practices tying the recruit to the rebel group. Richards shows how young recruits, often children abducted from their families and communities, were placed in a setting that echoed some of the characteristics of the society bush. The initiation of rebel recruits took place in the secluded bush, cut off from society (in this case the rebel camp) and behavioral rules and codes were introduced to recruits through the shared initiation experience, which in the case of the RUF tended to be violent.[36] Susan Shepler similarly draws parallels between rebel initiation and the secret society tradition and sees RUF markings of initiates through scarring as echoing the circumcision that takes place in the Poro or Bondo bush.[37]

Although tactical advantage and recruitment capacity were major concerns for both sides, the conceptual context into which this was inserted provided explanations much thicker than a simple analysis of technical skill and hardware support, which fed the imagination of those on the battlefield and their behavior when faced with their enemy. Stories of RUF rebels avoiding direct encounters with Kamajors were certainly rife both during and after the war—behavior supposedly fuelled both by fear of superior Kamajor military skill and spiritual power. Some of these stories are undoubtedly myths propagated by the Kamajors, and as we saw earlier, they at times also avoided direct or prolonged confrontation with the RUF as in Koribundu, but the pervasive belief in Kamajor power that has been notable also after the end of the war suggests that not all were attempts to bolster the confidence of one's own troops through mythmaking.[38]

The belief in the potency of magic and rituals resonated more broadly. Another interesting anecdote from the war is the account of the first meeting between the Nigerian ECOMOG commander Maxwell Khobe and the Kamajor leadership at Base Zero. It illustrates the ambiguous relationship between magic and power and the difficulty, and perhaps pointlessness, of challenging the worldview that has given rise to it.

Khobe was invited to Talia with some of his men to reinforce the CDF-ECOMOG relationship. Once at Talia, Hinga Norman offered to demonstrate the Kamajors spiritual powers to the commander and invited Khobe to have his body guard shoot at the High Priest Allieu Kondewa to show his imperviousness to bullets. Khobe refused, at which stage a chicken, immunized by the High Priest, was offered up as an alternative. No longer inhibited by the potentially deadly outcome, Khobe allowed his man to fire at the chicken from a short distance of about five meters. He could not hit the chicken and ended up emptying his entire magazine at the bird without harming it.[39]

Regardless of the reasons behind the body guard failing to kill the chicken at Talia that day, the suggestive power of the ritual protection was significant, both on the part of the Kamajors and most likely on that of the Commander. According to the eyewitness giving the account of the event, it allegedly changed the Kamajor-ECOMOG relationship dramatically. He suggested that the display of both the Kamajor organization, at Talia and through the initiation machine, and their alleged power and protection, was part of the reason why Khobe decided to keep the Kamajors from becoming too strong by limiting their weapons supplies.[40] Although a strong Kamajor force would put up an even greater challenge to the rebels, it would also challenge illicit mining activities pursued by ECOMOG troops and threaten the source of income it generated.[41] This strength perceived by Khobe could well be said to be derived from the cohesion of the group and *their* shared beliefs and practices, rather than the ECOMOG commander's shared understanding of them. In fact, an offer to test bullet immunity is a difficult one to accept from an ally—accepting the challenge and killing the high priest would undoubtedly have had devastating effects on the cooperation of the two forces, and accepting and failing would have significantly shifted the power balance between the two men.[42] Declining and instead accepting a lesser challenge allows the mystery (and by extension power balance) to remain intact. That said, it is also presumptuous to assume that the commander did not share the Kamajors' beliefs in the potency of the protection. As Ellis and Ter Haar have demonstrated in their work, the level of actual belief in supernatural power even amongst the elite (political and military) should not be underestimated or brushed aside.[43]

What is clear, however, is that the worldview in which the High Priest's immunizations were believable was the one shared by most of the actors engaged in the war. As such it held currency that could be translated into military advantage and even political leverage for community and logistical support.

Magic and Military Success?

Without question the belief in the spiritual power of the fighters impacted directly on the behavior of the Kamajors and their track record in battle. It is, however, less clear if this impact was predominantly positive or negative in terms of what we would see as a traditional approach to success.

The positive effects of the incorporation of ritual and a spiritual dimension stem from their close connection with the habitus of those involved—the wider cosmological context as just explored—tying into the explanatory frameworks engaged by the wider population for events taking place around them. In this sense, a fighter walking into battle and through a barrage of bullets without a scratch to his body is seen as the result of his magical protection. This belief is fed by the wider belief in the power of medicines and rituals in the community at large, known to be employed also by the fighters who in addition are believed to have particularly spiritually powerful initiators.[44] The fighters are then but a version of the secret societies seen throughout the country, and their power a form of the same known to civilians from their own initiation to *Poro* or *Bondo*. An account by a Kamajor ex-combatant illustrates this:

> I led an attack on AFRC forces on the main road between the diamond area in Kono and Makeni. The AFRC and the RUF attacked our position with heavy artillery. To prove that our society's medicine really works, I decided to go onto the road and face these guys. They were shooting at me, but I walked on towards them, and none of their bullets or artillery hit me. At a bridge there, I shot the driver of a rebel vehicle. There were six rebels in the truck. I captured the vehicle with all the ammunition inside. I drove it back

> to our base. It was one of the key battles because it weakened the rebels in the northern area.[45]

In this case, regardless of the actual causal chain of events, the interpretation dictated by the fighter's (and most likely his enemy's) beliefs has a positive outcome in that it reinforces beliefs. That is, even if he was not hit by the enemy's fire because they were not aiming properly or their weapons had malfunctioned, he interpreted his survival as the result of the power of his medicine.[46] The AFRC/RUF fighters who saw him walk straight through the salvo of fire and capture one of their vehicles, would likely have assumed a similar interpretation—partly because the alternative would have left them looking incompetent, but also because they shared the wider belief in the potency of spiritual protection.

The negative impact, from the perspective of how a Western observer would expect fighters to act on the battlefield, stems from the overconfidence brought on by the belief in one's invincibility. This is largely the view taken by Richard Iron in his expert witness report to the Special Court for Sierra Leone already quoted. Soldiers convinced of their immunity to bullets or the magical direction of the battle by controllers, would take risks that could compromise the success of the unit. The fatalistic attitude brought on by the holding of these beliefs was noted by British military staff engaged in the training of fighters to be incorporated in the new Sierra Leone Army (the RSLAF) after the war.[47] But the fickleness of spiritually induced success also calls into question its tactical benefits. The success of the "power" is dependent, both from a believer's and a nonbeliever's perspectives, on a series of factors that come back to the individual fighter's competence. A believing commander would have to rely on his fighters' observance of the behavioral rules laid down, to ensure his protection is not made void but also on the relative spiritual advantage of his forces as over the enemy. A nonbeliever would have to weigh the advantages to be had from greater courage and the intimidation of the enemy against its cost in terms of excess confidence undermining more classical soldiers' virtues such as weapons skills and coordination.

The nature of fighting during the war helped keep the beliefs and practices intact. As explored above, the relative lack of pitched battles and the emphasis on display over accurate targeting reinforced beliefs in the nonmaterial forces guiding events. The fact that so much of the ammunition went high above the heads of fighters—as a result of AK-47s being fired from the hip or held high in the air, dispersing bullets in all directions but straight ahead—allowed fighters to move through a battlefield heavy with fire without being hit by bullets. By the same token, death or injury could come at unexpected locations during battle from straying bullets and fire, giving the appearance that the bullets sought out offenders of the behavioral code. If, however, the aim of battle was not necessarily to kill as many of the enemy as possible but rather to force their retreat and lay claim to their weapons, success had less to do with traditional military efficiency than with effectively outperforming the enemy in terms of the display of power.

In this sense the shared beliefs of combatants and the context in which they operated colluded to create a situation where magic and ritual could indeed aid military success, primarily in terms of providing a strong sense of motivation to the fighters and in extension the intimidation of the enemy.

Effective Magic?

The elephant in the room when talking about the rational or logical application of magic is whether this application is in fact effective, especially relative to other elements of tactical behavior. It is the first question asked by any western observer when faced with the proposition that the use of magic can be both a functional and logical choice. Voltaire said that "it is unquestionable that certain words and ceremonies will effectually destroy a flock of sheep, if administered with a sufficient portion of arsenic" and we must tackle a similar skepticism when looking at the use of magic in war.[48] Effectiveness suggests that a given practice achieves what it sets out to do. The use of magic in the case of the Kamajors was intended to serve a number of interlinked and multilayered functions. As part of initiation it was intended to provide fighters with immunity to enemy weapons, directly impacting on their ability to survive battle. It was also intended to increase the confidence of fighters by instilling knowledge that enemy bullets would not hurt them, motivating them to go into battle without fear. As part of fighting, it was intended to provide fighters not only with immunity and fearlessness, but also with proactive powers to help them gain advantage over the enemy. Finally, magical powers were used to ensure discipline through its association with taboos.

Did the practices actually achieve these ends? While withholding judgment on whether the magic actually worked, it would appear as though the rituals and magic did have some positive effects relative to the desired outcomes. As a result of a combination of physical and psychological effects of the rituals they underwent, Kamajor fighters did go into battle with less fear. Potions with medical qualities—generally of a confidence boosting and pain reducing variety—combined with the promised powers of immunity, invisibility, or ability to control objects and people, allowed the fighters to approach combat with greater buoyancy. There were also additional related functions that were not explicitly the aim of the practices. Although group cohesion was a desired end of initiation, it was not explicitly tied to the use of magic. As Chief Norman said in his testimony, the exercise of initiation and the shared experiences of initiates were not too different from other training camps around the world and the use of magic was only one part of this exercise.[49] However, it seems to have provided a powerful finale to the experience through the communal sharing of magical endowment and its associated taboos.

On the battlefield, the magical powers possessed by the Kamajors also led them to evaluate events in a way that allowed them to maintain a degree of optimism that would perhaps have been lost on an outsider. The example of the Kamajor attack on Koribundu (related above) is a good example of this. Despite the lack of a decisive battle awarding the town to the advancing Kamajor fighters it was nonetheless seen as a success and victory, in part because of the powers assumed to be working on the side of the Kamajor soldiers in addition to their military activities. Although the ability to interpret events to their own psychological advantage was not an expressed aim behind the use of magic, it was a tangible effect. This also holds for the interpretation of deaths on the battlefield as resulting from broken taboos rather than ineffective magic. It was not an explicit aim of the taboos but an effective function derived from them. Similarly, the interpretation of events by the enemy was also affected by the shared cosmological space. The appearance of Kamajor fighters with their amulets and charms as well as particular instruments of war such as the RPG-bats and specially assigned fighters like the controllers sent a signal of magical prowess to the enemy—as we have seen this could have a tangible effect both on the enemy and the Kamajor fighters themselves.

Extrapolating from these areas of effectiveness it seems clear that the use of magic had its greatest impact on a tactical level. The strategic aims of the Kamajors were not expressed or conceived of in supernatural terms; in fact they were explicitly pragmatic; on the highest level it was about countering the rebel advance and providing protection to civilians in their local communities. This was subsequently translated into a struggle to restore the elected government after the 1997 coup as this was seen as the best route back to peace and an end to violence. Even the strategies employed in the theatre of operations were of a "universal" kind in that it focused on dislodging rebels from key areas and cutting off their supply routes to step by step bring them to their knees. However, on the tactical level the means employed to achieve these aims involved a high degree of magic. Returning to the framework borrowed by Bourdieu this makes good sense for two reasons: First, it was a logical result of the context in which the fighters emerged as an organized civil defense force (in so far as they shared a belief in the power of magic). Second, for exactly this reason, it served several functions that allowed the fighters to approach battle with greater courage and a sense of purpose.

But beyond this more tangible area of effectiveness, it also seems likely that the cosmology of fighters impacted on their way of war more generally, particularly the relatively nonconfrontational style of combat. Although several additional factors no doubt combined to shape this style of warfare, where pitched battles between opposing sides were not only rare but often actively avoided, cosmology has added some crucial imperatives to the mix.[50] Primarily it has contributed to a worldview where the unseen is considered to be as rich in agency as the seen. This translated into interpretations of events not solely dependent on visible physical attributes. For example, in the same way as it is not the aim of the man behind the rifle that is considered exclusively responsible for whether the bullet hits its target, battles are not assumed to be necessarily decided by physical force alone. As a result, both rebels and civil defense fighters would often make more noise than damage when encountering each other. With civilians they would, however, be far more ruthless. In part this is no doubt because they presented easier targets and were guardians of vital resources such as food and vehicles, but in part also because they were seen as part of a larger picture where information and knowledge passed through communities to influence tactical outcomes significantly. This is no different to other models of intelligence where the civilian population becomes a conduit for information of tactical and strategic importance. But as with the other areas where cosmology has an impact, in Sierra Leone, intelligence gathering would often be perceived as aided by supernatural power—particularly by the opposing side. As a result, cosmology and practices in part derived from it impacted on the general context within which tactical decisions were made and therefore affected also the general shape of warfare. In some ways this also made it effective.

Part III

CHAPTER 10

LOCAL PRACTICE AND THE WIDER WORLD OF WARFARE

Belief influences behavior and gives rise to practice. It feeds into a logic of behavior that is manifested through practice and that practice, without a clear understanding of the logic, can confuse conclusions about the behavior. Practice imbued with magic has suffered from exactly such confusion. But as we have seen, there is a logic that underpins the practice and allows it to be functional. In the case of Kamajor initiation practice, it is the culmination of a process informed by a set of beliefs and a worldview that incorporates active and accessible spiritual power. In a situation where power becomes of primary importance to protect individual and community life—as in war—the desire to access power in its broadest and most potent form increases. If practice that allows you to benefit from spiritual power (magic) is perceived as potent it is only logical that it should be used by those waging war. The incorporation of spiritual beliefs in warfare can therefore be expected wherever such beliefs have become part of the general disposition or commonsense world—what Bourdieu would call the habitus-capital-field nexus. In most of Africa, traditional religion and religion more broadly is generally part of that nexus. Wartime practice drawing on these beliefs has therefore been plentiful across the continent.

But this appeal to the resources at the disposal of a community, be they material or nonmaterial, is neither specifically Sierra Leonean nor even African. In any community faced with a challenge, the shared assumptions and beliefs in that community will shape the choices they make and the actions they take. Taking these influences into account requires a concerted effort to identify them as some of them may not be obvious to an outsider at first glance. The approach borrowed from Bourdieu allows us do this systematically by unpacking the "matrix of perceptions, appreciations, and actions" that shapes practice, whether the subject of study is Sierra Leone or Iraq.

Individuals and groups carry with them experiences, norms, perceptions, and beliefs that shape the way they develop practice. Understanding the ways in which these factors interrelate and combine, both with each other and with external pressures and developments, gives us greater insight into the relative significance of these influences as well as the nature of the practices they influence—in this case in war. Some of these influences may at first seem irrelevant or be misinterpreted as having a greater impact than is actually the case. Drawing on the insights gained from looking at the Kamajors in Sierra Leone and more cursorily at other wars in Africa, there are obvious implications for at least two areas beyond this context: (1) the general analysis of cultural factors in warfare and (2) practical engagement with conflicts in a culturally alien setting, whether through military intervention or civilian efforts to achieve peace or development.

Analyzing Culture in Warfare

As soon as a new conflict erupts there is a scramble amongst analysts to assess the motivations and dynamic of the war to either forecast its outcome or inform public and private sector policy toward it. Inevitably, the lack of detailed information skews much of this analysis in the early days of war. It takes time to identify the right questions to ask and even longer to identify the people who are in a position to answer those questions with insight. In that vein, it took many months before the genocidal dynamic of the Rwanda war became broadly represented in analysis. For a long time it was described as yet another ethnic war where all turned against all in pretty much equal proportion, as opposed to a civil war with defined actors where one side engaged in a particularly extensive campaign targeting ethnic Tutsi for extermination. Similarly, much analysis of the war that began in Afghanistan in 2001 still conflates all anti-NATO fighting groups as "Taliban," despite persistent evidence of the fragmented nature of the numerous insurgents that include both groups with overtly religious agendas and others with local tribal or economic ones. Seven years on from the outbreak of war this nuance is emerging, but seven years is a long time to wait.[1] Undoubtedly both of these cases suffered from a combination of lacking resources (analysts with the right linguistic and cultural expertise that could engage with the relevant actors to solicit the necessary information) and external political agendas (it is not always the case that external analysts and observers are after the objective truth or real life nuance, especially if analysis is undertaken to support the actions of one of the parties involved in conflict, for example in national media that feels an obligation to support the efforts of its armed forces regardless of circumstances or by policy analysts seeking to avoid certain action such as forced intervention under the Genocide Convention). They would also have suffered from compartmentalized analysis in the sense that those who really did understand the nuances of the situation were likely trying in vein to communicate their knowledge from academic or local institutions that did not have the ear of either policy makers of mainstream analysis and media outlets. In both cases the costs in terms of human life and general destruction have been very significant indeed. Whether these costs could have been limited if a more solid understanding of the dynamics of the conflicts had been sought and found at an earlier stage is up for debate, but it certainly seems likely that improved "visibility" would undo some of the debilitating fog of war, whether for a military or civilian engagement.

Before any kind of practical engagement is undertaken there should be good analysis. If we can use the toolkit derived from Bourdieu's theory of practice to make sense of the Kamajors' wartime practices in Sierra Leone, we should be able to apply it to similar effect elsewhere. Although the conclusions in relation to the Kamajors are unlikely to translate directly to other armed groups, the tools that helped us reach them should be able to help situate influencing factors in very different cultural settings as well. Rather than conclude that all human activity is directed by a neat means-ends relationship, with equally neat constituent parts, Bourdieu's approach makes room for complex choices in part made up by rational calculations and in part formed by a "common-sense" at least partially subconscious and formed by numerous experiences and norms not always taken into account actively by the agents involved. But just as it avoids a view of human activity that is strictly instrumentalist in its interpretation, neither does it succumb to relativist interpretations where "cultural" choices are left unquestioned. Instead it strikes a balance between the two by recognizing the internal logic of practice formation with both its rational calculations and cultural and context-specific influences—some conscious and some subconscious. For war, and strategic theory, this approach allows us to unpack the logic of why certain means are thought to maximize outcomes and are therefore chosen. Continuing along this path, we can place religious beliefs (including those that incorporate belief in the power of magic) under Bourdieu's heading habitus—the shared common sense of the group or community. This common sense is constituted by norms, beliefs, and generally shared perceptions that have been formed through direct and indirect (historical) experience and interaction. Some will be dependent on events, others on physical realities, but together they form and are continuously formed by the individuals that share the norms. The habitus in turn interacts with the context (field) in which the group in question exists. In peacetime this will be influenced by socioeconomic realities as well as political developments near and far. Depending on the particular locality and its position within local, regional and global developments the relative weight of these influences will differ. In wartime, these factors are usually realigned to place violent politics in its midst. In Sierra Leone it meant that whereas in peacetime the context was in part defined by poverty, in part by weak government authority and capacity (and depending on the particular locality, the ability of local authorities to govern well, local economic opportunities and so on), when the war arrived this was superseded, if not entirely displaced, by the threat posed by the rebel invasion. Finally, we can fit the tools at the disposal of the group or community under Bourdieu's heading capital. On a basic level this encapsulates physical resources such as weapons and people. But it also includes conceptual tools and patterns of behavior that are utilized or have the potential to be utilized to address a particular concern, or in Bourdieu's words interest. In war, the main concern of the protagonists tends to relate to the capture of power (to be able to compel an opponent to bend to one's will) or strategic resources through the use of force. Both aggressors and defensive responses need to mobilize this kind of capital to be able to achieve their strategic objectives, be they to replace or maintain the status quo. Practice emerges at the nexus of habitus, field, capital, and interest.

The habitus-field-capital-interest matrix from which particular practices emerge says as much about processes of human behavior as it does about the behavior in specific cases. The habitus will differ depending on location and can be limited to a group within a location and only shared in part by the wider population, even in that locality. For example, a particular church congregation in a local English town will share some understandings of the world with the rest of the town inhabitants but not others that will be specific to their religious beliefs and the social interactions that take place within the congregation. We may in fact have to investigate numerous and overlapping habitus to come to an accurate understanding of the dynamic at play. Similarly, the field in which these dispositions that constitute the habitus are being formed provides another clue to the way practice or activity might develop. Finally, the resources (capital) available to the group will limit, or enable, them in their pursuits. Unpacking these influences and dynamics will reveal their logic and enable a more nuanced understanding of practices that may at first seem baffling. The tools for doing this therefore serve both as tools for achieving deeper understanding and as a bit of a checklist that can guard against some of the more common but easily avoidable traps of simplification and distortion. On closer inspection much of it is the systematic application of common sense.

So what is this practically applicable "toolkit?" To understand the motivation behind a particular activity, or even to anticipate the likely behavior of a group, we must understand the combined inputs of the group's motivation, make-up, and operating environment. These are all enabling (or restraining) factors impacting on strategy and tactics. All three are impacted by the habitus, field, and capital of the group that inform its interests and strategies and, in extension, specific tactics.

MOTIVATION

To understand what motivates a group we need to consider the group's interests based on the interaction between habitus and field. In practice this tends to translate into a review of the worldview of its members (based on stated aims, actual track record and the identity of members), the socio-economic context from which they emerged and in which they operate, and pressures they are under. In other words, what aspects of their reality have been threatened or compromised and how are they seeking to address this? Cosmology is one part of this puzzle. Although it does not necessarily carry more weight than the other parts it must still be considered to complete the picture. The main challenge in this case is to differentiate between the worldview propagated through rhetoric and the worldview actually shared by group members that influences and drives behavior. This requires consideration not only of rhetorical statements but of the history of the group, its constituencies, and the challenges that led to the upheaval that resulted in them opting for armed struggle. In particular, the group's own understanding of those challenges, as opposed to the way they are perceived by an external observer, are likely to betray their motivation more accurately.

Also important for an understanding of motivation is the appreciation of the social relationships that existed before the war and how those relationships have been transformed in times of war. Have these simply been displaced by new relationships or have existing ones been brought into the new setting? Would the relationships that existed prior to conflict lend themselves to wartime organization as an armed group? In many African wars, patronage relationships have aided mobilization, either by replacing previous patron-client relationships that were perceived as dysfunctional or by patrons calling on existing clients to take up arms. Hoffman makes this point in his work on West Africa and argues that many of the fighting groups were not new or distinctly military organizations but presented the militarization of existing patronage relationships. Similarly, the patron-client relationships that are created in war tend to remain also in peace, placing the dynamics of many armed groups on a continuum where the transition between war and peace is less absolute than processes of for example demobilization would suggest. In those circumstances war is less of a corruption of peacetime functions and relationships than an extension of them under different conditions.[2] These relationships will help explain the motivation of a group in so far as they will give guidance as to where the impetus for action is coming from—whether it is a top-down imperative or a rebellious one that challenges authority, not only at a national level but also locally. It is also likely to give insight into the risk of resumed violence at the end of a war—are fighters genuinely committed to the renouncement of violence or are they simply following orders? If the latter, how committed are those giving the orders and are they able to recall their clients to resume armed activity should this commitment falter?

As much as the worldview influences explicit agendas it will influence specific relationships and the motivation of a group will almost always be determined by a combination of both.

STRATEGY/TACTICS

A group's strategy will inevitably be closely related to its motivation and main aim. The strategy is after all the path that guides you toward your desired end. The Kamajors' strategy appears to have been significantly less influenced by their cosmology than their tactics were, but this by no means guarantees that this same distinction applies to all armed groups. The investigation into the nature of strategy and tactics should therefore be guided by questions about the interrelationship between a group's motivation and desired aim and the constraints placed on their means for achieving this by beliefs and norms held by the fighters as well as the resources (capital) available to them to pursue it.

To understand this interplay we have to look at what they have done in light of what we know about their motivation and worldview. What they have done is likely to at least in part betray their capability (resources/capital) but also strategic and tactical choices. For example, as seen in Sierra Leone, an attack that on the surface looks like a poorly coordinated and executed attempt to claim territory by defeating the enemy in battle, may in fact have been intended to intimidate the enemy and persuade them to retreat without suffering a debilitating defeat in terms of casualties. However, to see the attack in that light one would have to consider the social fabric, relationships and beliefs of the fighting groups and how this acts as a constraint on certain types of strategies (set battles aimed at inflicting large-scale casualties) whilst encouraging others (attacks based as much on appearance and suggestions of power as the execution of these).[3] Going through this same exercise for another group may therefore show that an attack (or some other component to strategy) that seems to serve a particular purpose is in fact intended to serve a different one that only becomes clear if the group's motivation and understanding of the world is appreciated.

Tactics are decided by a combination of available means, the group members' understanding of behavior (what behavior yields which results), and the nature of the enemy. This is where cosmology or other culturally specific beliefs and practices are most likely to have an impact as it feeds into the conscious as well as subconscious understanding of the specific resources available for fighters. In Sierra Leone this ranged from modern weaponry and drugs to ritual protection and patron-client relationships. In other places, different capital may take this place, be they beliefs and perceptions about social organization, tribe or ethnicity, gender or a manifest destiny. How a group fights is very likely to say a great deal about what informs their tactics *if* placed in the context of their motivation and worldview. It is also likely to say something about the direction of fighting, particularly where certain types of actions are considered taboo, particularly brave or honorable.

Viability

The viability of a group, or their likely success/survival, is dependent on their ability to replenish supplies and manpower and maintain morale. Some groups manage to stay operational for decades whereas others collapse within months. This is in large part dependent on the group's ability to secure weapons and keep its fighters from being killed, but it is also decided by its ability to mobilize support and attract new fighters. For any armed group fighting an insurgency, or counterinsurgency, the support of the local population is crucial. This can be achieved through intimidation or by winning genuine proactive support for the cause. Determining the group's ability to do either requires an understanding of the way the group is perceived by the wider population as well as how the group is attempting to influence this perception. A key consideration is how the group claims legitimacy and to what extent this resonates with the population. As both the Sierra Leone example and other African wars have shown, worldview often plays a significant role here. In a society where religious belief is widespread and used broadly to explain the world and its machinations, this will be a logical point of reference for mobilizing support, albeit almost always in combination with other concerns. In other places, ideological conviction or different languages of rights can play the same role. Belief systems are also likely to impact on the ability of a group to maintain internal discipline and therefore its general cohesion. Understanding how loyalty is encouraged and how effectively, is a key building block for understanding the likely ability of a group to survive or grow stronger.

The ability to recruit new fighters is another fundamental indicator for a group's viability. In groups where casualties are great or regular (e.g., where suicide attacks are a key tactic) this becomes a more pressing concern. New recruits can be secured either by appealing to potential new members to join of their own free will or by coercing people into the ranks. The former is clearly much more closely aligned to the legitimizing rhetoric of the group, and an understanding of how far, and in what ways, this rhetoric resonates with the likely recruitment pool will give a good indication of their likely success. Assessing this rhetoric vis-à-vis the habitus and capital of the recruitment pool (whether the population as a whole or a particular segment) may in turn unveil crucial dimensions to the group's recruitment ability that otherwise would have remained obscured. This was certainly the case with the Kamajors where the appeal of protecting the community was powerfully complemented by the offer of both magical protection and a legitimizing myth giving the civil defense effort spiritual impetus as well as support.[4]

Providing the nonhuman capital for a group is also crucial to its survival. Weapons and food have to be sourced for a group to wage war. Extras, such as sophisticated communications equipment or means of transport are of varying importance in different contexts. The majority of Kamajor fighters used their own bodies both to communicate and move across distances, reducing their reliance on supply chains. In their case they also sourced a significant amount of weapons from defeated enemies and food from the relatively bountiful forests they operated in, further increasing their independence from external support. At the same time, they nonetheless relied on external support to acquire the heavier weapons needed to win larger-scale battles like those fought to retake the Tonga diamond fields in 1997–1998. They also relied on supplies from villages and towns—sometimes willingly given and other times not. But although fear and brutality undoubtedly played into their ability to secure supplies from the civilian population this mixed with respect and awe at their perceived power. These dynamics will vary for all groups but are likely to be influenced at least in part by their beliefs (religious, ideological, cultural) and agendas as these will affect their ability to secure the support of suppliers.

Intervention and the Context Trap

A prime motivation for good analysis is effective engagement. Externally driven intervention in wars comes in a variety of forms and with a wide range of mandates. However, all share a common challenge in finding a workable and effective mode of engagement. Often, differing perceptions, values, and social organization pose obstacles to interveners. For military intervention this is perhaps particularly challenging as military institutions are generally more resistant to changing doctrines or strategic thinking. As a result, most military interventions look very similar. At a basic level they all involve an overall strategy of deploying confrontational force to bring about a desired outcome—generally an end to hostilities and the establishment of sympathetic governments. Although some involve large-scale deployments and others a smaller number of specialist forces, it is very rare that the local context is considered much beyond the geographic terrain and what is known about the enemy's weapons capability. The culturally sensitive dimension to counterinsurgency operations almost always comes later and with greater difficultly, and notably once the initial intervention has proved unable to achieve the desired results.[5] Forms of this type of intervention have been seen in contexts as varied as Sierra Leone, Kosovo, Afghanistan, and Iraq.

The outcomes have, however, been quite different. Some interventions have met with significant success whereas others have fuelled an escalation in conflict. In the four cases just cited, except for Kosovo, the expectations were largely opposite to the actual outcome. In Sierra Leone, the British intervention in 2000 was undoubtedly intended to halt the rebel advance to allow for the effective evacuation of foreign nationals and, if possible, force them back to the negotiating table in a weaker state. But in the end it had a major impact on the momentum of the war as a whole and bore significant responsibility for the complete end to hostilities.[6] In Afghanistan and Iraq on the other hand, the interventions in 2001 and 2003 unleashed bitter insurgencies. In Iraq this was predicted by many observers, notably those with a deeper understanding of the internal dynamics in the country and the brewing sectarian conflicts bubbling away under the authoritarian lid of the Ba'athist regime. In Sierra Leone, the outcome was not as vociferously predicted, but nonetheless makes sense if we consider the conflict through the lens suggested earlier. This way, we can see how even a small number of British soldiers would provide a shock to the system of war that was in play. In a theatre of war where both sides generally avoided pitched battles and the highest number of deaths was among civilians, a new party focusing on killing a large number of rebels was not only not conforming to the ruling convention of war but also undermining the will of many of the targeted actors to risk continued fighting. Previous engagements by the private military contractors Executive Outcomes had a similar effect and as a result the rebels specifically demanded their removal from the country as part of their negotiations in Abidjan in 1996. The British, as a state actor, were, however, harder to sideline than a private company that had both financial concerns and a somewhat dubious international legal position to consider.

It is worth pausing to consider the relative success of Executive Outcomes in 1995–1996. The company drew its staff primarily from amongst former apartheid-era South African special forces. Most of their men had extensive experience supporting insurgents in Angola and Mozambique and would have been familiar both with the nature of fighting forces based on small units with strong patronage-style lines of command and the types of supernatural beliefs that often were part of the mix. Coupled with the fact that a large number of Angolan and Mozambican veterans from these same wars were also members of staff they undoubtedly had an advantage in terms of their ability to make sense of the environment they were entering. This is not to say that Angola or Mozambique were the same as Sierra Leone, but the dynamics would have resembled each other more closely than they did Northern Ireland or the Falklands. But like the British, the private contractors also had a military strategy that involved using deadly force in a targeted way, focusing firepower on the rebels. Executive Outcomes' direct use of Kamajors for intelligence gathering and to infiltrate rebel camps in advance of attacks allowed them to benefit from local expertise whilst also challenging the ruling convention of war by launching large-scale attacks with heavy weapons and air support. In short they both drew advantage from the local dynamic and challenged it.

So why could private military contractors and the British be so relatively successful in Sierra Leone when a similar strategy of a short, targeted use of force has been so problematic both in Iraq and Afghanistan? The answer is obviously that the contexts were very different, which is exactly what lies at the heart of the argument made here. The dynamics at play among "neo-Taliban" groups in southeastern Afghanistan or within the al-Qaeda affiliated Islamic State of Iraq are clearly very different from those of the RUF in Sierra Leone, both in terms of motivation, strategy, and tactics. And it is the specific circumstance of a group that decides how an intervention against it will play out. Strategies and tactics therefore need to be informed by deep knowledge of the particular group and setting, not the last encounter with an insurgent group halfway across the globe or even next-door. So whereas what we learn from looking at the Kamajors in Sierra Leone is instructive in that it shows us the importance of understanding the role of magic in shaping their tactical behavior it is not the case that all other armed groups will be similarly influenced. The ability to differentiate between approaches that allow us to evaluate practices in different settings and ones that extrapolate specific conclusions from one setting to another is central to informing effective strategy.

There are clearly also nonviolent practical applications of a culturally informed approach to behavior in warfare. In addition to informing engagement with an opponent on a battlefield it can inform strategy off the battlefield. Public diplomacy campaigns and psychological operations hinge on connecting with the minds of the people it targets, be it the opponent or the civilian population. This is clearly already a key concern in Iraq and Afghanistan but is becoming ever more pressing for Africa as well, especially as renewed interest both by military and private investment actors is gaining considerable speed. The launch of the US Africa Command (Africom) will require a significant public diplomacy effort to allow it to become more than a vehicle for occasional military intervention. Its advocates insist it has a much broader mandate to support security and policing capacity and to assist and prepare African allies not only for war, but also for humanitarian crisis response and other civic duties. It is the first command to include integrated staff from the State Department and USAID. To fulfill its mandate, Africom will need to cultivate the capability to understand local contexts and how best to connect with the people in them. In the African setting, an understanding of the role of religious beliefs and practices both in everyday life and with specific reference to military activity will be as crucial as understanding other culturally specific norms and behavior.

This same understanding can also be a tool for psychological operations (psyops) or information warfare. More closely connected to the military sphere, psyops are specifically focused on influencing an opponent or other target groups' beliefs, values or behavior. In the US Department of Defense's own dictionary description it is "planned operations to convey selected information and indicators to foreign audiences to influence their emotions, motives, objective reasoning, and, ultimately, the behaviour of foreign governments, organizations, groups, and individuals."[7] Less of a dialogue than public diplomacy it in fact requires an even greater understanding of the target group to correctly assess how they will respond to different techniques. In any kind of information or propaganda war the information distributed needs to be tailored to the target audience to ensure that it achieves its intended aims. Just like advertising agencies spend vast resources to assess how to best connect with different sets of customers to sell a particular product, selling ideas also needs to identify the right connection points. Without a thorough understanding of the ways in which the target audience understands the world and interprets behavior this will be very hard to achieve. The Rhodesian government of Ian Smith came to this realization well into the country's civil war when they began to distribute leaflets denouncing the spirits the insurgents claimed to be collaborating with but also articulated the government's agenda with reference to the spirits of ancestors—not pitting belief in spirits against the lack of such belief but spirit against spirit. In Mozambique, the Naprama countermovement to the Renamo rebels also built its case around a spiritual rhetoric that challenged that of the rebels.[8]

However, whether the right answer is to adopt the exact language of the opponent when attempting to win the hearts and minds of the wider population, or to co-opt members of the enemy's ranks, will depend on the specific context and what the intervener is trying to achieve. Lying to gain the support of local populations is very unlikely to lead to lasting gains. In the case of the Smith government in Rhodesia, the attempt to affect people's behavior by attempting to tap into their view of the world largely backfired as few were convinced that a government that had previously called them superstitious and gullible had performed a 180 degree turn and come around to appreciate the power of spirits. But without understanding what they think, feel, and believe, no language is likely to resonate constructively. Much social psychology that is applied to warfare in western states is derived from studies of behavior in Nazi Germany and although there are undoubtedly some universal truths about human behavior there are also, as this study has attempted to show, many types of behavior that are distinct to a locality.[9] To persuade a group of people to change their view or allegiance, that specific group's needs must be understood. But they must be understood without falling into the traps of either purely culturalist theories (which risk generating materials of persuasion similar to those distributed by the Smith government) or purely instrumentalist explanations that assume the drivers of conflict and behavior are essentially the same as those in the intervening country and any appearance that seems to suggest otherwise is mere chimera.

This is true also for engagements aimed at peace negotiations. Externally imposed negotiation agendas have collapsed many peace processes because they failed to connect with the actual concerns of the groups involved and therefore could not gain their support and commitment in the long run. The multiple peace initiatives in Sudan's Darfur region have been a recent example where the nature, let alone agendas, of the many armed factions have been elusive to mediators. As a result, identifying the issues that need to be addressed has been very difficult and the armed groups have had no reason to trust that a process unable to correctly assess their grievances will be able to offer any sustainable ways of addressing them.

Without doubt, the additional resources allocated to intelligence analysts in terms of language training and field exposure will go some way toward addressing these concerns. But the principal insurance against poor analysis—and in extension practical engagement—is asking the right people the right questions. Both can prove tricky in an unknown setting. Doing it successfully requires not just an ability to ask about culture or religion, but an ability to identify what Bourdieu called "intentionality without intention," or rational action not necessarily based on reason. It is the subconscious commonsense world that shapes behavior in the most subtle, and therefore often most powerful, ways. Looking for the interplay of dispositions, perceptions, and pressures we can hopefully engage more meaningfully with cultural dynamics in and attributes to warfare and move beyond simplistic characterizations and conclusions, be they of a new barbarism or a clash of civilizations, or contemporary concerns with global jihad. By doing so, we can hopefully improve not only our understanding of particular groups or phenomena but also the more general process of practice formation and behavior in war.

Notes

1 Introduction

1. Hinga Norman quoted in Martin Wollacott, "Sierra Leone: Special Report," *The Guardian,* June 6, 2000.
2. Taylor was eventually extradited to the Special Court for Sierra Leone on March 29, 2006 and has since been transferred to The Hague to stand trial.
3. Hinga Norman had been the national coordinator of the Sierra Leone Civil Defense Forces, and the deputy minister of defense. At the time of his indictment, he was minister of the Interior.
4. Francis Fukuyama put forward the end of history thesis to explain the victory of liberal democracy at the end of the cold war, suggesting that the big ideological battle for supremacy had now been resolved in the West's favor because liberal democracy coupled with capitalism was the natural endpoint of political development. Francis Fukuyama, *The End of History and the Last Man* (New York: Free Press, 1992).
5. Robert Kaplan, "The Coming Anarchy," *The Atlantic Monthly*, February 1994.
6. Authors like Mary Kaldor, Mark Duffield, and David Keen tried to do this by offering explanations resting on expediency, seeing the type of violence as reflective of the limited means available and a desire to nonetheless maximize gain in this environment. Mary Kaldor, *New and Old Wars: Organized Violence in a Global Era* (Cambridge: Polity, 1999); Mark Duffield, "Post-Modern Conflict: Warlords, Post-Adjustment States and Private Protection," *Civil Wars*, Spring 1/1 (1998); David Keen, *The Economic Functions of Violence in Civil Wars*. Adelphi Paper No. 320 (London: Oxford University Press for the International Institute of Strategic Studies, 1998); William Reno, *Warlord Politics and African States* (Boulder, CO: Lynne Rienner, 1998).
7. Adda Bozeman, "War and the Clash of Ideas," *Strategic Intelligence and Statecraft. Selected Essays* (Washington DC: Brassey's [US], 1992); Peter J. Katzenstein, *Cultural Norms and National Security: Police and Military in Postwar Japan* (Ithaca, NY: Cornell University Press, 1996); Colin Gray, *Modern Strategy* (Oxford: Oxford University Press, 1999); John Jandora, "War, Culture and the Interpretation of History: The Vietnam War Reconsidered," *Small Wars and Insurgencies*, 9/2 (1998); Ann Fitz-Gerald, "Understanding Local Dynamics in Civil Wars," *Journal of Civil Wars*, 3/1 (2000).

8. Eva Evers Rosander and David Westerlund (eds.), *African Islam and Islam in Africa: Encounters between Sufis and Islamists* (London: Hurst, 1997).
9. Paul J. Smith, "Transnational Terrorism and the al Qaeda Model: Confronting New Realities," *Parameters* (Summer 2002), pp. 33–46; Lauren Ploch, "Africa Command: U.S. Strategic Interests and the Role of the U.S. Military in Africa," CRS Report for Congress, updated March 10, 2008; Daniel Byman, "Remaking Alliances for the War on Terrorism," *Journal of Strategic Studies*, 29/5 (2006), 767–811. The space devoted to Islamic groups in the Patterns of Global Terrorism reports published by the Office of the Coordinator of Counterterrorism in the U.S. State Department significantly increased from 2001.
10. Kondewa was found guilty by the Special Court for Sierra Leone, initially in August 2007 and later by the Appeals court in May 2008.
11. Witness testimony, SC-SL May 15, 2006, afternoon session, Trial chamber 1, transcript not yet available. Similar testimony was offered by the first accused Chief Sam Hinga Norman (the National Coordinator for the CDF) SC-SL January 24–February 7, 2006, Trial chamber 1.
12. There are notable exceptions, although in the spirit of anthropological and historical study these works have focused primarily on telling a story and detailing behavior rather than extrapolating lessons for practical engagement and even less so military engagement. Stephen L. Weigert, *Traditional Religion and Guerrilla Warfare in Modern Africa* (London: Macmillan, 1995); David Lan, *Guns & Rain: Guerrillas & Spirit Mediums in Zimbabwe* (London: J. Currey, 1985); Heike Behrend, *Alice Lakwena & the Holy Spirits: War in Northern Uganda, 1985–97* (Oxford: J. Currey, 1999); Stephen Ellis, *The Mask of Anarchy: The Destruction of Liberia and the Religious Dimension of an African Civil War* (London: Hurst, 1999).

2 "Magic" in Contemporary Africa

1. Although it is important to note that in medieval Europe magic was not a clear-cut conceptual arena either, and the nature of supernatural, or "unnatural" power was sometimes considered as ambiguous as in the African context. Witchdoctors or the "cunning folk" often treaded a fine line between white and black magic, and they were often accused, at some stage in their career, of witchcraft. However, going in the other direction, accused witches could buy time as well as prestige by offering their services as witch finders. As in Africa, the occupational space at the edge of science and the border of religion offered cures for illnesses—some of which worked and some of which did not—and explanations for phenomena for which no other explanations were available. At times these were welcomed by society and at times condemned. See Robin Briggs, *Witches and Neighbours: The Social and Cultural Context of European Witchcraft*, Second edn. (Oxford: Blackwell, 2002).
2. Much of the press coverage is in relation to the destructive side of the practice—ritual killings, curses, or hostile spells—and is reported with

varying degrees of skepticism. For example, reports of a man being transformed into a goat on public transport in Sierra Leone kept a skeptical tone but nonetheless spread the news of the alleged event. In instances of ritual killings, the tone of reporting tends to remain neutral with a focus on the beliefs of people involved.

3. While some churches have opposed progressive agendas such as universal suffrage, others have embraced and pushed such agendas, e.g., through the development of liberation theology in Latin America.
4. There are varying views on what exactly these external challenges are, and consequently what responses they generate, but in general the opening up of economic activity through increasingly diverse international trade, the impact of political conditionality by international financial institutions on domestic policy, and the results of increasing global communications speed and information exchange, can be seen as indicative of the types of factors envisaged.
5. In many cases other religious practices, notably Christian and Muslim, have also stepped in to fill this need—often in combination with traditional practice. Parallels could perhaps also be drawn with the resurgence of fundamentalist strands of Islam in the Middle East and Asia in the past couple of decades, also in response to both rapid social change and a perceived lack of tangible reward. See e.g., Mary Ann Tétreault and Robert Allen Denemark, *Gods, Guns, and Globalization: Religious Radicalism and International Political Economy* (Boulder, CO: Lynne Rienner, 2004); Nilüfer Göle, "Snapshots of Islamic Modernities," *Daedalus,* 129/1 (Winter 2000).
6. Peter Geschiere has separated the two purposes of witchcraft in Cameroon as "levelling" and "accumulation"—both tied to the distribution of wealth. Peter Geschiere, *The Modernity of Witchcraft: Politics and the Occult in Postcolonial Africa* (Charlottesville, VA: University Press of Virginia, 1997) 1–25.
7. Adeline Masquelier, "The Invention of Anti-Tradition: Dodo Spirits in Southern Niger," in Heike Behrend (ed.), *Spirit Possession, Modernity and Power in Africa,* 34–50 (Oxford: J. Currey, 1999).
8. For political ideology to serve this function, it generally has to be holistic and offer interpretations and solutions for all aspects of life. The totalitarian political movements of the twentieth century did this explicitly and in many cases competed directly with religious institutions for the hearts and minds of the populations.
9. Weigert, *Traditional Religion and Guerrilla Warfare.*
10. Although not all insurgents are Salafi or even Sunni, for al-Qaeda in the Islamic Maghreb and al-Qaeda in Iraq their local agendas are framed specifically in Salafi thought.
11. In the strict sense of the word, cosmology refers to the study of the universe but has very different connotations depending on whether that study is undertaken by an astrophysicist, a philosopher or metaphysicist, or a religious scholar. Religious cosmologies tend to be focused on creation myths.

However, the way in which I use it here is closer to the philosopher's or metaphysicist's conception of cosmology as the totality of space, time and all phenomena contained by the two. Of course the creation myth and the spiritual world becomes part of this whole and often direct the understanding of the other constituent parts of it where supernatural belief is widespread among a population.

12. The films coming out of Nigeria's "Nollywood" are particularly prone to storylines centered on the abuse of spiritual power by the villains, eventually defeated through the power of Islam or Christianity. But despite the victorious "new" religion, the power of the old practices is described in matter of fact terms as a given part of society.
13. The Azande live in what is today the border region between Sudan, Central African Republic, and the Democratic Republic of Congo. Pritchard undertook his studies in the area in the 1920s and published his findings in *Witchcraft, Oracles and Magic among the Azande* in 1937. E. E. Evans-Pritchard and Eva Gillies, *Witchcraft, Oracles, and Magic among the Azande* (Oxford: Clarendon Press, 1976).
14. Evans-Pritchard and Gillies, *Witchcraft, Oracles,* pp. 22–23.
15. See e.g., A. Ashforth, "An Epidemic of Witchcraft? The Implications of Aids for the Post-Apartheid State," *African Studies,* 61/1 (July 2002); J. Stadler, "Rumor, Gossip and Blame: Implications for HIV/AIDS Prevention in the South African Lowveld," *AIDS Education and Prevention,* 15/4 (August 2003).
16. The Yanga and Simba clubs were each fined $500.
17. Heike Behrend has made a categorization similar to Geschiere's accumulation and leveling that conceives of spirit power in northern Uganda as used for "healing" and "killing" (or both) in the civil war. Heike Behrend, "Power to Heal, Power to Kill, Spirit Possession and War in Northern Uganda (1986–1994)," in Heike Behrend (ed.), *Spirit Possession, Modernity and Power in Africa,* 20–33 (Oxford: J. Currey, 1999).
18. Geschiere, *The Modernity of Witchcraft,* p. 96.
19. Behrend, "Power to Heal, Power to Kill," p. 22.
20. This school of thought was pioneered by the "Manchester School," made up of writers such as Max Gluckman, Victor Turner, Clyde Mitchell, and Max Marwick.
21. J. D. Krige, "The Social Function of Witchcraft," in Max Marwick (ed.), *Witchcraft and Sorcery* (London: Penguin Books, 1982) p. 274.
22. Ibid.
23. Mary Douglas has compared witchcraft to infectious disease to illustrate this power of the hidden: "Infection and occult harm are both hidden from observation: a carrier can transmit disease to others without showing any signs of infection: a witch looks like anyone else. From their hiddenness both forms of harm afford the same kind of opportunity for accusation and exclusion." Mary Douglas, *Risk and Blame: Essays in Cultural Theory* (London: Routledge, 1992) p. 83.

24. See e.g., *What Future? Street Children in the Democratic Republic of Congo* (Human Rights Watch: 2006).
25. The Joint Revolutionary Council accused the NDDC of witchcraft in a statement sent to the newspaper *Vanguard* in August 2008, subsequently published by the paper on August 13, 2008.
26. Ibid.
27. Stephen Ellis and Gerrie Ter Haar, *Worlds of Power: Religious Thought and Political Practice in Africa* (London: Hurst, 2004) p. 2.

3 SPIRITS IN BATTLE

1. Behrend, *Alice Lakwena & the Holy Spirits.*
2. Taylor withdrew his lawsuit after it became clear that he would have to appear in court in London. Taylor objected to Ellis's writings in his 1999 book *The Mask of Anarchy.*
3. Ellis, *The Mask of Anarchy, A Journey without Map* (Camerapix, 2003), James Brabazon (dir.).
4. We shall return to this in greater detail in part II when we explore the specific case of the Kamajors during the Sierra Leone civil war.
5. "Eastern Congo Ravaged: Killing Civilians and Silencing Protests," Vol. 12 No.3 (A) (Human Rights Watch, 2000).
6. K. B. Wilson, "Cults of Violence and Counter-Violence in Mozambique," *Journal of Southern African Studies*, 18/3 (1992), p. 544.
7. By Holy Spirit Movements, plural, is meant the HSM and HSMF of Alice Lakwena, as well as the successor movements (the Lord's Army, the United Holy Salvation Army (UHSA), the United Democratic Christian Army (UDCA), and the Lord's Resistance Army (LRA). However, the most important are the HSM/HSMF and the LRA, although the UHSA and UDCA were in effect the LRA by a different name in its first years.
8. The Acholi are a Nilotic people who speak Acholi or Lwo resident primarily in northern Uganda, in a region also referred to as Acholiland.
9. Luwero is the area in the south where some of the worst atrocities were committed during the war between the Uganda National Liberation Army (UNLA) that dislodged Idi Amin in 1979, and the National Resistance Army (NRA) of Yoweri Museveni that took power in 1985, and the perpetration of which was associated with the Acholi because of their prominent presence in the UNLA.
10. Lakwena, means "messenger" in Acholi but is also the name of the chief spirit that possessed Alice Auma.
11. Behrend, "Power to Heal, Power to Kill," p. 24. In this case, witchcraft was seen as the destructive practices inspired by bad spirits and increasingly polarized by the Christian framework that was used by the HSM.
12. Personal communication, civil society mediator engaged with the LRA, London 2004. In an on-the-record meeting at the Royal Institute of International Affairs in London, the Archbishop John B Odama of the

Acholi Religious Leaders' Peace Initiatives (who has been present at a number of meetings with the LRA) expressed his reluctance to voice an agenda on behalf of the rebels. But he also said that in meetings a desire to come out of the bush as long as the security of the rebel group leader Joseph Kony and the LRA could be guaranteed had been expressed (referring particularly to the meetings held with Kony by minister for the North Betty Bigombe in 1994) suggesting that survival and safety are now the overriding concerns for the rebels. "Northern Uganda: Africa's Forgotten Conflict," Royal Institute of International Affairs. London, February 3, 2004. This has been borne out to some degree by on-off peace talks in South Sudan since the summer of 2006 where the key issues on the agenda have been amnesty and a security guarantee for returning rebels.

13. Sverker Finnström, "Living with Bad Surroundings: War and Existential Uncertainty in Acholiland, Northern Uganda," PhD (Uppsala University, 2003) p. 281.
14. Ibid.
15. Ibid. Finnström discusses the possible origins and truth of this rumor and the people involved in some detail in his thesis.
16. This curse came in the form of a condemnation of the LRA by a prominent elder in Gulu, who according to Finnström displayed his penis and said, "if these children who are in the bush originate from my penis, I curse them"—the gravest curse known in Acholi. Ibid., p. 284.
17. Clearly these 140,000 must have incorporated also more ancient ancestors and sympathizers as the death toll, although high, did not reach such numbers at the time and would in any case not have made such a numerous force immediately available to the HSMF when they set out on their campaign.
18. Behrend, *Alice Lakwena & the Holy Spirits,* p. 48.
19. Ngwabi Bhebe and T. O. Ranger, *Soldiers in Zimbabwe's Liberation War* (London/Harare: J. Currey, 1995); Ngwabi Bhebe and T. O. Ranger, *Society in Zimbabwe's Liberation War* (Oxford: J. Currey, 1996); T. O. Ranger, *Peasant Consciousness and Guerrilla War in Zimbabwe: A Comparative Study* (London: J. Currey, 1985); Norma J. Kriger, *Zimbabwe's Guerrilla War: Peasant Voices* (Cambridge: Cambridge University Press, 1992).
20. Lan, *Guns & Rain,* p. 4.
21. See Behrend, *Alice Lakwena & the Holy Spirits,* p. 38.
22. J. Gordon Melton and Martin Baumann (eds.), *Religions of the World: A Comprehensive Encyclopaedia of Beliefs and Practices* (Santa Barbara, CA: ABC Clio, 2002) gives the religious breakdown for Mozambique as: Ethnoreligionism (indigenous beliefs) 50.4 percent, Christian 38.4 percent, Muslim 10.5 percent, for 2000.
23. Weigert, *Traditional Religion and Guerrilla Warfare,* pp. 75–76.
24. Ibid., p. 76.
25. It is, however, worth remembering the actual level of fraud documented (in addition to the genuine beliefs held) within Renamo, more on which later.

26. Wilson, "Cults of Violence and Counter-Violence," p. 549. Harry West recounts similar tales of Frelimo "sorcery" and alliances with mediums in Harry G. West, "Creative Destruction and Sorcery of Construction: Power, Hope and Suspicion in Post-War Mozambique," *Cahiers d'études Africaines* XXXVIII (3)/147 (1997) p. 685.
27. Weigert, *Traditional Religion and Guerrilla Warfare,* p. 94.
28. Lan, *Guns & Rain,* p. 196.
29. Broadcast by Ian Smith, January 18, 1973, quoted Ibid., p. 157.
30. Behrend, "Power to Heal, Power to Kill," p. 27; Behrend, *Alice Lakwena & the Holy Spirits,* p. 51.
31. Behrend, "Power to Heal, Power to Kill," p. 30. Interestingly, the Holy Spirit is said to have left Kony in 2001, and will reappear in a new medium in time. But Kony is nonetheless thought to remain powerful as a result of his years with the spirit and the wisdom and knowledge imparted to him.
32. Behrend, *Alice Lakwena & the Holy Spirits,* p. 47.
33. "Abducted and Abused: Renewed Conflict in Northern Uganda" (New York: Human Rights Watch, 2003) p. 21.
34. Alex Vines, *Renamo: From Terrorism to Democracy in Mozambique?* 2nd rev. edn. (London: J. Currey, 1996) p. 112 quoting João Fabião, a former Renamo combatant. Weigert, *Traditional Religion and Guerrilla Warfare,* p. 80 and Wilson, "Cults of Violence and Counter-Violence," p. 544 also tell of the varying degrees of ritual protection received by the different ranks.
35. Wilson, "Cults of Violence and Counter-Violence," p. 545.
36. Ellis, *The Mask of Anarchy,* p. 119.
37. Although many of these live bullets in fact hit their targets, the beliefs of the fighters were strong enough to accept this as a sign of the impurity of the fighter that was injured rather than as proof of a lack of magical power.
38. Behrend, *Alice Lakwena & the Holy Spirits* and Wilson, "Cults of Violence and Counter-Violence," p. 545.
39. Behrend, *Alice Lakwena & the Holy Spirits,* p. 61.
40. Perhaps the response from the government was more interesting than the offer itself. In a government press briefing following the offer, minister for defense Ruth Nankabirwa stated that the government was considering the plan suggested by Nacotha, saying "[t]here are some factors the Government has been overlooking in this war, and Kony's alleged use of spiritual powers is one of them . . . We are therefore considering adopting this method in addition to the existing ones." She was flanked and backed up by James Nsaba Buturo, minister for information. However, the response was subsequently criticized and withdrawn. Quoted in Richard Mutumba, "Government Plots Spiritual War on LRA," *The New Vision,* July 18, 2003. See also "Uganda's Spiritual War Heats Up," *BBC News Online,* July 22, 2003.
41. See Weigert, *Traditional Religion and Guerrilla Warfare,* p. 80, Vines, *Renamo,* chapter 4.
42. As already mentioned, the idea of immunization failing only when a counter spell has been cast, or when the soldier has broken the rules set to him,

comes again in northern Uganda, Liberia, and Zimbabwe. In addition, in many cases effective counter spells can be cast only by someone close to the target for such a spell, that is, from within the own camp. Behrend, *Alice Lakwena & the Holy Spirits,* p. 108; Lan, *Guns & Rain,* p. 37; Wilson, "Cults of Violence and Counter-Violence," p. 543.

43. Weigert, *Traditional Religion and Guerrilla Warfare,* p. 81.
44. Ellis, *The Mask of Anarchy,* pp. 11–12.
45. Indeed the stories go back further, particularly to the reign of President Samuel Doe in the 1980s, as detailed by Ellis in his research. But the stories continue to the present day and the current conflict between Charles Taylor's government and the LURD. Ibid., U.S. Department of State, "Country Reports on Human Rights Practices: Liberia," (Washington DC: U.S. Department of State, 1993–2001); James Brabazon, "Liberia: Liberians United for Reconciliation and Democracy (LURD)" (The Royal Institute of International Affairs: Armed Non-State Actors Project, Briefing Paper No. 1, 2003); and *A Journey without Map,* Brabazon (dir.) (video documentary following the LURD in 2002).
46. Stephen Ellis, "Liberia 1989–1994: A Study of Ethnic and Spiritual Violence," *African Affairs,* 94 (1995) p. 192.
47. *A Journey without Map,* Brabazon (dir.).
48. Quoted in Ellis, *The Mask of Anarchy,* pp. 146–148.
49. Denis Tull, "The Reconfiguration of Political Order in Africa: A Case Study of North Kivu," Hamburg African Studies (Hamburg: RFA Institut für Afrika-Kunde Hamburg, 2005), p. 239.
50. Quoted in "DRC: Mai Mai leader Gedeon of Manono Territory-Known 'Good Guy,' Accused Cannibal," United Nations Organization Mission in the Democratic Republic of Congo (MONUC), April 14, 2004.
51. Behrend, *Alice Lakwena & the Holy Spirits,* p. 58.

4 Thinking about Practice in Warfare

1. Academia has however recently seen a growing interest in a serious assessment of the role of religion in conflict, not least because of the rise of radical Islam and the ongoing religious dimension to the conflicts in the Middle East and Central Asia. But wider study of the role of religion in conflict is also starting to receive greater attention, a development this book is intended to contribute toward.
2. And strategy here refers to the whole range of strategic activity, from the grand strategy level to operational tactics: in effect, the thinking and analysis that directs activity in warfare.
3. Bertrand Russell, *Human Society in Ethics and Politics* (London: Allen and Unwin, 1954) p. viii.
4. J. Garnett, "Strategic Studies and Its Assumptions," in K. Booth J. Baylis, J. Garnett, P. Williams (ed.), *Contemporary Strategy: Theories and Policies* (London: Croon Helm, 1975) p. 17.

5. Ibid.
6. M. L. R. Smith, *Fighting for Ireland?: The Military Strategy of the Irish Republican Movement* (London: Routledge, 1995) p. 4, emphasis added.
7. Katzenstein, *Cultural Norms and National Security,* p. 19.
8. Theo Farrell, "Culture and Military Power," *Review of International Studies* 24 (1998) p. 415.
9. The theory of practice is of course only part of the work done by Bourdieu. The intention here is to help structure the argument and articulate a framework for approaching the assessment of the role of magic in African warfare, rather than apply Bourdieu's work in its entirety to either the world of war or religion. To do this Bourdieu's concepts as outlined here are useful analytical tools. And it is as such that they are included.
10. Pierre Bourdieu, *The Logic of Practice* (Cambridge: Polity, 1990) pp.123, 62.
11. Pierre Bourdieu, *In Other Words: Essays Towards a Reflexive Sociology* (Oxford: Polity, 1990) p. 123, and p. 10.
12. Pierre Bourdieu, *Outline of a Theory of Practice* (Cambridge: Cambridge University Press, 1977) p. 261.
13. Bourdieu, *The Logic of Practice,* p. 58.
14. Michael C. Williams, *Culture and Security: Symbolic Power and the Politics of International Security* (London: Routledge, 2007), Chapter 2.
15. Reprinted in Bourdieu, *The Logic of Practice,* p. 41.
16. Ibid., p. 10.
17. Ibid., p. 11.
18. Whether the belief is genuine is a topic that lends itself to endless discussion as there will always be individuals who profess not to share the belief and at the same time others who do. In Sierra Leone the general "climate" of belief—as outlined in chapter 6—is one of genuine conviction and subscription to the practical implications of correctly administered ritual practice.
19. Secret societies and the wider cosmology of the Kamajors will also be explored in greater detail in chapter 6.
20. They may of course still exist and perform a similar function, but the reference points would be different, and as a result so would the "form" of practice exercised by the society.
21. Bourdieu, *The Logic of Practice,* p. 10.
22. Williams, *Culture and Security.*
23. Ibid., Chapter 2.

6 THE SIERRA LEONE CIVIL WAR AND CIVIL DEFENSE

1. The origins and course of the Sierra Leone civil war have been extensively documented by others and this is not intended as an exhaustive overview of events. For that, the work of David Keen, Paul Richards, and Lansana Gberie provide excellent guides. The focus here will be on the civil defense militias and the outline of the war is intended as a backdrop to that and must therefore acknowledge its limitations.

2. Paul Richards, *Fighting for the Rain Forest: War, Youth & Resources in Sierra Leone* (African Issues; Oxford: J. Currey, 1996); David Keen, *Conflict and Collusion in Sierra Leone* (Oxford: J. Currey, 2005); Lansana Gberie, *A Dirty War in West Africa: The RUF and the Destruction of Sierra Leone* (London: Hurst, 2005); J. L. Hirsch, *Sierra Leone: Diamonds and the Struggle for Democracy* (Boulder, CO: Lynne Rienner, 1996); Adekeye Adebajo, *Building Peace in West Africa: Liberia, Sierra Leone, and Guinea-Bissau* (International Peace Academy Occasional Paper Series; London, Boulder, CO: Lynne Rienner, 2002).
3. Reginald Cline-Cole, "Perspectives from yet Other Places, Spaces and Voices: A Commentary on Michael Watts' 'Development and Governmentality,'" *Singapore Journal of Tropical Geography,* 24/1 (2003), pp. 40–41.
4. Reno, *Warlord Politics and African States,* p. 114.
5. Keen, *Conflict and Collusion in Sierra Leone*, p. 37.
6. The Economic Community of West African States Monitoring Group (ECOMOG) was deployed by ECOWAS to Liberia in 1990 in response to the outbreak of war in December 1989.
7. Estimated numbers as given by Krijn Peters, *Re-Examining Voluntarism: Youth Combatants in Sierra Leone*, Monograph, 100 (Pretoria: Institute for Security Studies, 2004) p. 9.
8. Keen, *Conflict and Collusion in Sierra Leone*, p. 94.
9. Ibid., Chapter 6.
10. For an overview and discussion of the peace initiatives up to 1999, see David Lord, *Paying the Price: The Sierra Leone Peace Process*, ed. Catherine Barnes (Accord: An International Review of Peace Initiatives; London: Conciliation Resources, 2000).
11. It is widely assumed that the RUF had been collaborating with elements of the AFRC before the coup and had infiltrated Freetown in May 1997 in anticipation of the officers striking. Keen, *Conflict and Collusion in Sierra Leone*, Gberie, *A Dirty War in West Africa*.
12. The Conakry Agreement, October 23, 1997.
13. As the United Nations Assistance Mission in Sierra Leone (UNAMSIL).
14. "Operation Palliser" was launched on May 7, 2000 and came to a close on June 15, 2000. During the operation, Sierra Leonean troops Foday Sankoh was captured and imprisoned.
15. The British-led International Military Assistance and Training Team were established in 2002 and were still present in the country in 2008.
16. By this time Taylor had gained the Liberian presidency but was facing a renewed rebellion by a group calling themselves Liberians United for Reconciliation and Democracy (LURD), allegedly launched from and supported by Guinea.
17. Kabbah declared the war over at a peace ceremony on January 18, 2002.
18. Dr. Lavalie formed the Eastern Region Defense Force together with Joe Demby (who later was appointed vice president after 1996 elections). Joe Demby testified to this effect to the Special Court for Sierra Leone, SC-SL,

Witness Alfred Joe Demby, February 9, 2006 as did several former Kamajors I spoke to in Kenema, Bo, and Freetown. See also Mariane C. Ferme and Danny Hoffman, "Hunter Militias and the International Human Rights Discourse in Sierra Leone and Beyond," *Africa Today,* June 2004, p. 75.

19. Interview, Kamajor senior commander, Koribundu February 2005 and Kamajor commander Bo, February 2005.
20. Michael Jackson, *In Sierra Leone* (Durham, NC: Duke University Press, 2005) p. 143.
21. The 37,000 CDF fighters suggests a massive recruitment took place as an estimate from 1996, just as the Kamajors were starting to assert themselves more effectively, numbers them at only 2,500. Patrick Muana, "The Kamajoi Militia: Civil War, Internal Displacement and the Politics if Counter-Insurgency," *Africa Development*, XXII/3/4 (1997) p. 90. Although there was a dramatic increase in recruitment in 1996 and 1997, disarmament figures tend to be artificially inflated by the procedures for demobilization and disarmament; anyone with a recognized weapon can disarm and qualify for the demobilization package often leading to noncombatant relatives and dependents of combatants registering for demobilization with surplus weapons as well as actual combatants (or even taking the place of actual combatants at the behest of a commander looking out for his clients to the detriment of others). On the other hand, many of the weapons used by the Kamajors, notably single barreled shotguns were not recognized for DDR and excluded many Kamajor combatants from the process. Numbers quoted by former Kamajors themselves go far beyond the DDR figures with claims of a total membership of close to 100,000 fighters. On balance, the truth is probably closer to the DDR figure, although the core of fighters would have been smaller, with many of the disarmed "combatants" serving in capacities not directly involving combat.
22. Muana, "The Kamajoi Militia," p. 83.
23. Ibid.
24. Interview, Kamajor commander Bo, February 2005.
25. "Nature of the Conflict," *Witness to Truth: Report of the Sierra Leone Truth and Reconciliation Commission* (Vol. 3a) pp. 105–106.
26. There are several variations on the narrative of the start of initiation in Bonthe and the emergence of the Kamajors as a magically empowered force. These and their significance are explored in greater detail in chapter 7.
27. Alison L. Smith, Catherine Gambette, and Thomas Longley, "Conflict Mapping in Sierra Leone. Violations of International Humanitarian Law from 1991 to 2002. Preliminary Edition," p. 44. However, the centralized institution The Civil Defense Forces / Sierra Leone would only become a legal entity in 1997, and truly centralized after the return of Kabbah to power in 1998.
28. This perception was likely reinforced after the death of the Ghurkha Security Guards' commander, which many believed to be the result of information passed from the SLA unit they were fighting alongside to the RUF. Gberie, *A Dirty War in West Africa*, p. 90.

29. Interview, former EO employees Freetown February 2005, and Kamajor ex-combatant, Freetown, February 2005. Even after EO left, the Kamajors and CDF kept contacts with the company and its successor Sandline and continued collaborations into 1998.
30. Letter from Johnny Paul Koroma to ECOWAS, August 1997. Quoted in "Sierra Leone: Time for a New Military and Political Strategy. Africa Report N° 28" (Brussels: International Crisis Group, 2001).
31. Hinga Norman had, according to his testimony, been in the Royal Signals of Sierra Leone from 18 years of age and later served with the British Army of the Rhine in Germany in 1960, the UN peacekeeping mission to Congo, and received training at Mons Officer Cadet School in Aldershot, England. Testimony by Sam Hinga Norman to the Special Court for Sierra Leone, Trial chamber 1, January 24, 2006, p. 46.
32. The War Council was formed in November 1997 at a meeting in Talia (Base Zero).
33. Ferme and Hoffman, "Hunter Militias and the International Human Rights Discourse," p. 76.
34. Danny Hoffman, "The Meaning of Militia: Understanding the Civil Defence Forces of Sierra Leone," *African Affairs,* 106/425 (2007) pp. 639–662.
35. Interview, senior Kamajor commander, Kenema, May 2005.
36. SC-SL Witness Alfred Joe Demby, February 15, 2006, p. 12.
37. Interview, senior Kamajor commander, Kenema, May 2005.
38. Danny Hoffman, "Expert Report on the Kamajors of Sierra Leone; Prepared for the Fofana Defense Team" (Special Court for Sierra Leone, 2006) p. 28.
39. The ULIMO faction fighting against Charles Taylor in neighboring Liberia's civil war had originated from Liberian refugees in Sierra Leone and not only shared a common interest with the CDF in countering the Taylor-sponsored RUF, but many also came from the border region and had strong family links to the Kamajor fighters.
40. Maxwell Khobe was the Nigerian head of the ECOMOG forces in Sierra Leone. There were a number of Liberians in the Kamajor ranks who had previous combat experience in the Liberian civil war and joined the ranks of the CDF. Most of these operated out of Base One because of its proximity to Liberia; very few visited Base Zero. Many of the Liberian CDF fighters were also former ULIMO fighters. When the second phase of the Liberian civil war commenced in the late 1990s, many of these fighters joined LURD to seek the overthrow of Charles Taylor. Hoffman, "Expert Report on the Kamajors of Sierra Leone," p. 29.
41. In an SLBC broadcast on May 25, 1997, the AFRC announced that "all Kamajors are to be disbanded forthwith. No more Kamajors, no more civil defence forces as from now. We are the National Army." Quoted in Ibrahim Abdullah, "Bush Path to Destruction," *Africa Development*, XXII/3/4 (1997) p. 231.

42. Interview, Kamajor initiators, Bo, May 2005, Kamajor ex-combatant Freetown, May 2005, and Kamajor senior commander Kenema, May 2005.
43. Although generally only minimal training was given to recruits who picked up their skills "on the job." Interviews, Kamajor ex-combatant and commanders.
44. Kondewa's name can also be spelled Kundorwai, Kondowah or Kondewah, and he also went by the title "King" Allieu Kondewa.
45. Interview, Kamajor ex-combatant Freetown, February 2005, initiators Bo, May 2005, and Kamajor senior commander Kenema, May 2005.
46. *Witness to Truth: Report of the Sierra Leone Truth and Reconciliation Commission,* p. 53. The initiators in turn had assistants and apprentices that aided them in their work and often became full-fledged initiators themselves with time. E.g., Mama Munda's apprentice is now the national spokesperson for initiators. Interview, Kamajor initiators, Bo May 2005.
47. Kabbah had retained the office of Minister of Defense for himself, a point repeatedly made by the defense teams for the CDF accused at the Special Court for Sierra Leone arguing for the inclusion of the president in the recognized and actual chain of command, and as a result ultimately responsible in the command-control chain.
48. Hoffman pays a great deal of detail to Iron's report in his own statement and in part constructs his argument as a critique of Iron. It is perhaps important to note that Iron appeared as a witness for the prosecution, which has argued for a clear chain of command within the CDF that leaves the indictees responsible for atrocities carried out by fighters. Hoffman, on the other hand, provided his report and subsequently testimony for the Fofana defense. Hoffman has, however, spent many years studying and personally observing the Kamajors, including during the war years, whereas Iron came in as a military rather than Sierra Leone witness to conduct much briefer research to formulate a view on command-control in the CDF.
49. Hoffman, "Expert Report on the Kamajors of Sierra Leone."
50. Richard Iron, "Military Expert Witness Report on the Civil Defence Forces of Sierra Leone" (Office of the Prosecution: Special Court for Sierra Leone, 2006); this was also the main contention of the prosecution at the Special Court.
51. Interview, senior Kamajor commander, Kenema, May 2005.
52. Danny Hoffman makes this point very forcefully in his expert witness report to the Special Court. Hoffman, "Expert Report on the Kamajors of Sierra Leone."
53. Ibid., p. 30.
54. SC-SL, Witness TF2–005 (War Council member), February 16, 2005 (Closed session), p. 17.
55. SC-SL Witness TF2–222 (Kamajor ex-combatant), February 17, 2005, p. 92.
56. Interview, Kamajor commanders, Koribundu and Kenema, February and May 2005.

57. Famous in part because it signaled a turning point in the CDF campaign against the RUF/AFRC but also because it has been the focus of much of the prosecution case against the CDF in the Special Court and as such far better documented than many of the other operations that took place at the time.
58. SC-SL, Witness TF2–190 (Bobor Tucker, Kamajor senior commander), February 10, 2005, p. 46. Tucker was a generally well respected and regarded Kamajor commander who gave his testimony at the Special Court in February 2005. Tellingly of his reputation (and in part the testimony he gave) many former Kamajor commanders I spoke to did not denounce Tucker despite his appearance at the court as a prosecution witness. Most other inside witnesses have been denounced as liars and disgruntled criminals holding a grudge against the indictees.
59. Hoffman makes this point in his expert witness statement to the Special Court for Sierra Leone, and it was a point that was further reinforced by several of the commanders I spoke to. Hoffman, "Expert Report on the Kamajors of Sierra Leone."
60. Smith, Gambette, and Longley, "Conflict Mapping in Sierra Leone," p. 45.
61. Interview, Kamajor commanders and ex-combatant Koribundu, Bo and Freetown, February 2005.
62. SC-SL, TF2–190 (February 10, 2005) p. 42.
63. Interview, former Executive Outcomes personnel, Freetown, February 2005.
64. Thomas Legg and Robin Ibbs, "Report of the Sierra Leone Arms Investigation" (London: House of Commons, 1998), Paragraph 7.9. However, the CDF commanders I spoke to claimed that the weapons never reached them.
65. Interview, senior Kamajor commander, Kenema, May 2005.
66. Whether looting was admissible was a recurrent theme in cross-examinations at the Special Court and witnesses gave very different accounts of the Kamajor rules relating to looting and theft. It would appear that some commanders made an effort to prevent their fighters looting by referencing this rule whereas others did not.
67. Interviews, Kamajor ex-combatants Bo and Freetown, February 2005, SC-SL testimonies from Witnesses TF2–190 (February 10, 2005), TF2–222 (February 17, 2005).
68. Ferme and Hoffman, "Hunter Militias and the International Human Rights Discourse," p. 82.
69. Ibid., pp. 82–83 quoting from RSLCDF, "Recommended Values and Standards" (1999).
70. Interview, Kamajor commander, Koribundu, February 2005.
71. Hinga Norman announced the operation on a radio announcement broadcast by the BBC. SC-SL, Witness TF2–027, February 18, 2005, SC-SL, Witness TF2–005 (War Council member) February 16, 2005, p. 14.
72. Interview, senior Kamajor commander, Kenema, May 2005.

7 Spiritual Context as Habitus and Capital

1. Muslim 60 percent, indigenous beliefs 30 percent, Christian 10 percent. *CIA Factbook*, September 2008. Although these numbers are only an estimation, and many hold multiple faith identities (primarily overlapping Muslim or Christian with traditional faiths), it is clear that Sierra Leone is a primarily Muslim society but with a significant Christian population and lasting traditional belief systems. A census was undertaken in Sierra Leone in 2004, and although information on confessional affiliation was collected, the results have not been made public. Figures for population size and distribution are available from Statistics Sierra Leone, the official statistics bureau launched in 2003. The previous census was held in 1985.
2. The Mende are the largest ethnic group in Sierra Leone and comprise around 30 percent of the population, concentrated mainly in the south and east of the country.
3. K. L. Little, "The Function of 'Medicine' in Mende Society," *Man,* 48 (November 1948) p. 127.
4. Ibid.
5. There are exceptions to this, although they are rare, where women have been initiated into both Poro and Bondo, as is the case with mabole, women seen as both male and female, as detailed by Mariane C. Ferme, *The Underneath of Things: Violence, History and the Everyday in Sierra Leone* (Berkeley: University of California Press, 2001) pp. 74–79. Poro and Bondo reach beyond the Mende in Sierra Leone and can be found also in Liberia and Guinea, as can other similar societies that go by other names but function in a very similar manner.
6. Ibid. p. 76.
7. C. H. Bledsoe and K. M. Robey, "Arabic Literacy and Secrecy among the Mende of Sierra Leone," *Man, New Series,* 21/2 (June 1986) p. 205.
8. Ferme, *The Underneath of Things,* p. 180.
9. Interview, Mariama Conteh, West Africa programme manager, Conciliation Resources (Freetown, February 2005).
10. Paul Richards, "The Emotions at War: Atrocity as Piacular Rite in Sierra Leone." In Perri 6, Susannah Radstone, Corinne Squire, and Amal Treacher (ed.) *Public Emotions* (London: Palgrave Macmillan, 2006).
11. On the role of Masonic lodges and other secret societies of European origin in Africa, see Ellis and Haar, *Worlds of Power.* An interesting example of the migration of rituals in the opposite direction is the Warrior Institute at Morgan State University in Baltimore where young men are taught using the "Warrior Method" that is described as "an adaptation of the African Poro Society created to fit the realities of being a young Black male in the United States."
12. In the West the most common types of secret initiation society that occupies a shadow land on the edge of the socially acceptable and legal are overtly criminal or satanic in nature. Criminal networks that operate on

principles of secrecy and initiation tend not to have a spiritual dimension but emphasize loyalty and discipline; the Camorra in southern Italy or the Triads in Hong Kong and Macao would fit this bill as would many other organized crime groups across the world. Satanic sects are viewed as suspicious by the wider society in part because of an agenda that is perceived to be destructively egoistic and some of the means associated with it, particularly sacrificial rituals. Although there have been relatively few instances of documented human sacrifice in satanic sects, the association with an explicitly violent and self-serving ideology has pushed them outside the acceptable mainstream. For further consideration of criminal secret societies and their organizational structure and dynamic see B. H. Erickson, "Secret Societies and Social Structure," *Social Forces,* 60/1 (1981).

13. K. J. Beatty, *Human Leopards: An Account of the Trials of Human Leopards before the Special Commission Court* (London: H. Rees, 1915); D. Burrows, "The Human Leopard Society of Sierra Leone," *Journal of the Royal African Society* 13/50 (January 1914).
14. See Ellis, *The Mask of Anarchy* and Ellis and Haar, *Worlds of Power.*
15. Little, "The Function of 'Medicine' in Mende Society," pp. 127–128.
16. Little describes them as "a compound of herbs mixed as a rule with other natural ingredients, such as soil and leaves, and the whole is saturated with water. Medicines of this kind (saweisia) are usually employed ritually for 'washing' certain social offences and crimes, as well as for the cure of physical ailments. Other medicines, particularly the kind used forensically, may consist of a wide variety of miscellaneous objects, such as cowries, old razor blades, ribbon, feathers, animal fat, human nails etc. Often the medicine is tied up in a piece of cloth, or it may be contained in the horn of an animal, such as a sheep or a goat." Ibid., pp. 128–129.
17. Rosalind Shaw, "Cannibal Transformations: Colonialism and Commodification in the Sierra Leone Hinterland," in Henrietta L. Moore (ed.), *Magical Interpretations, Material Realities: Modernity, Witchcraft and the Occult in Postcolonial Africa* (London: Routledge, 2001) p. 66. "Once prepared, such medicines could be applied by eating or drinking them, washing with them, boiling them and inhaling the steam, rubbing them into incisions in the skin, dropping them in the eyes, wearing an undershirt soaked in them, sewing them into a belt tied around the waist, or enclosing them in an amulet... hung around the neck."
18. For an overview of the types of magic practiced with the help of the Qur'an and its written word, see B. A. Donaldson, "The Koran as Magic," *The Muslim World,* 27 (1937).
19. Bledsoe and Robey, "Arabic Literacy and Secrecy," p. 210, Shaw, "Cannibal Transformations," p. 59 and Little, "The Function of 'Medicine' in Mende Society," p. 129.
20. Donaldson, "The Koran as Magic," p. 255.
21. Bledsoe and Robey, "Arabic Literacy and Secrecy," p. 211.
22. Ibid., p. 210. Meanings in Mende culture are generally assumed to be multifaceted and often hidden from view, so that the literal meaning of a text

is only one layer of its meaning with more hidden between the lines. This is added to by the already complicated readings of the Qur'an as prescribed by *istikhara* and *fal*. For a detailed study of the role of meaning in Mende society see Ferme, *The Underneath of Things*.

23. Bledsoe and Robey give a list of uses for which mori magic is sought that includes; pursuing love affairs, curing barrenness, divining the future, gaining employment, winning court cases, harming enemies, passing school exams, defeating witches, and driving away demons of madness. Bledsoe and Robey, "Arabic Literacy and Secrecy," p. 210. Interviews with Kamajors and non-Kamajor civilians confirmed this range of activities by morimen as well as herbalists in peacetime as well as during the war.
24. Ferme, *The Underneath of Things*, p. 3.
25. Most famously by J-F, Bayart in Jean-François Bayart, *The State in Africa: The Politics of the Belly* (New York: Longman, 1993).
26. Interview, Kamajor senior initiator, Bo, May 2005.
27. Ritual killings by medicine men primarily for private clients have been reported from practically all countries on the continent. Particularly well documented have been instances in Nigeria and South Africa, possibly due to a combination of their advanced media and police.
28. Bourdieu, *Outline of a Theory of Practice*, Chapter 2.

8 Creating Magic Soldiers: Kamajor Mobilization and Initiation

1. Interviewee of David Caspar Fithen, "Diamonds and War in Sierra Leone: Cultural Strategies for Commercial Adaptation to Endemic Low-Intensity Conflict," PhD (UCL, 1999) p. 243.
2. Muana, "The Kamajoi Militia," Hoffman, "The Kamajors of Sierra Leone," p. 9.
3. According to one of Hoffman's interviewees, elephants and bush-cows were thought to have medicines in their stomachs and bladders. Hoffman, "The Kamajors of Sierra Leone," p. 52.
4. Ibid., p. 53.
5. Muana, "The Kamajoi Militia," p. 86.
6. Danny Hoffman, "Like Beasts in the Bush: Synonyms of Childhood and Youth in Sierra Leone," *Postcolonial Studies*, 6/3 (2003) p. 303. Hoffman also notes how Kamajor war songs "frequently linked fighting the enemy with the essence of manhood."
7. Account given to Muana in 1996, Muana, "The Kamajoi Militia," pp. 87–88.
8. Interview, Kamajor ex-combatant, Freetown, February 2005.
9. Interview, Kamajor initiator, Bo, May 2005.
10. As discussed in Chapters 2 and 5, the powers themselves could be used for both good and bad and the intent of the practitioner or spirits/person that bestowed the powers becomes central to legitimacy.
11. Interview, Kamajor senior initiator, Bo, May 2005.

12. Interview, Kamajor ex-combatants Freetown and Bo February 2005; Special Court Sierra Leone, transcript Court proceedings February 10, 2005, testimony given by Bobor Tucker, Kamajor ex-combatant and commander of the Death Squad, witness TF2–190.
13. Mama Munda Fortune was also illiterate, and claimed she received the necessary Qur'anic writings from other initiators skilled in mori magic, or from her patron bush devil (spirit) Kassela. Hoffman, "The Kamajors of Sierra Leone," p. 78.
14. Interview, Kamajor senior initiator, Bo, May 2005.
15. Interview, senior Kamajor commander, Kenema, May 2005.
16. Ex-Kamajor fighter quoted in Hoffman, "Like Beasts in the Bush," p. 304.
17. Interview 2001/07, male ex-combatant (RSLMF/CDF), December 2001, Blama. Krijn Peters, "The Storm Is Not yet Over? Interviews with Ex-Combatants from the War in Sierra Leone," (Unpublished typescript: Technology & Agrarian Development Group, Wageningen University & Research Centre, The Netherlands, 2002) p. 36.
18. Interview 2001/04, male ex-combatant (CDF), November 2001, Tissor. Ibid., p. 40.
19. "Ex-Combatant Views of the Truth and Reconciliation Commission and the Special Court in Sierra Leone," (Freetown: The Post-conflict Reintegration Initiative for Development and Empowerment, International Center for Transitional Justice, 2002) p. 9.
20. Interview 2001/08, male ex-combatant (CDF) December 2001, Blama. Peters, "The Storm Is Not yet Over?," p. 44.
21. "Youth, Poverty and Blood: The Lethal Legacy of West Africas Regional Warriors," (New York: Human Rights Watch, 2005) pp. 3, 11.
22. Interview, senior Kamajor commander, Kenema, May 2005.
23. As was the case with another Kamajor ex-combatant interviewed in Freetown February 2005.
24. Interviews, Kamajor commanders, Koribundu and Freetown February 2005, Kenema, May 2005.
25. Dyan Mazurana and Kristopher Carlson, "From Combat to Community: Women and Girls of Sierra Leone" (Hunt Alternatives Fund, Women Waging Peace, 2004) pp. 12–13.
26. Interviews, Kamajor senior commander Koribundu, and Kamajor ex-combatant Bo, both February 2005.
27. As seen in chapter 6, the widespread Bondo and Poro societies were gender-specific and the circumstances under which women could be initiated into the male Poro society were very unusual.
28. Interview, Kamajor ex-combatant Freetown, February 2005.
29. Interview, Kamajor ex-combatant Freetown, May 2005.
30. Interview, Kamajor ex-combatant Freetown, February 2005.
31. Peters, "The Storm Is Not yet Over?," p. 36.
32. Ferme and Hoffman, "Hunter Militias and the International Human Rights Discourse," p. 77. Le 50,000 at 2005s exchange rate equals approximately £10. By comparison a bag of rice (in peacetime) costs around Le 80,000.

33. Steven Archibald and Paul Richards, "Converts to Human Rights? Popular Debate about War and Justice in Rural Central Sierra Leone," *Africa,* 72/3 (2002) p. 355.
34. On the move between factions see "Youth, Poverty and Blood."
35. However, of the instances of abuse recorded by the Sierra Leone Truth and Reconciliation Commission, only 6 percent were attributed to the CDF (although the majority of these were attributed to the Kamajors) while the RUF accounted for 60.5 percent. TRC Report, Findings, p. 13. The CDF abuses peaked in 1997 correlating closely with the AFRC/RUF junta interregnum. "Statistical Appendix to the Report of the Truth and Reconciliation Commission of Sierra Leone," *Witness to Truth: Report of the Sierra Leone Truth and Reconciliation Commission* (Freetown: Sierra Leone Truth and Reconciliation Commission, 2005) p. 30.
36. Hoffman, "The Kamajors of Sierra Leone," p. 136.
37. Ibid., p. 138.
38. This standard was, however, as we shall see repeatedly disregarded when necessity outweighed the availability of suitable recruits.
39. A female ex-combatant interviewed in Mazurana and Carlson, "From Combat to Community," p. 13. Whether she—as a woman—proceeded to go through full initiation, the rituals she did go through clearly had an effect on her, one which would most likely have been only further entrenched by more elaborate initiation.
40. SC-SL, Witness TF2–021, November 2, 2004. The witness said he was captured and made to join the Kamajor society after having originally been captured by rebels in Kailahun. He was captured by the Kamajors during AFRC rule (1997–1998) and says he was taken to Talia and initiated together with approximately 400 other recruits. A "ronko" is a piece of clothing, like a vest, worn by Kamajors in battle. "Sassman" is a locally brewed alcohol, and a "rubber" the plastic container it is stored in.
41. Interview, Kamajor ex-combatant Freetown, February 2005.
42. Hoffman, "The Kamajors of Sierra Leone," p. 84. It is also an idea familiar from northern Uganda where one of the Holy Spirit Safety Precautions forbade fighters to turn back when going in to battle. See chapter 3.
43. Interview, civil society activist involved in conducting interviews with and collecting testimonies from Kamajor ex-combatants for the Truth and Reconciliation Commission, London May 2004 and Bo February 2005. Breasts, genitals, ears, and tongues were also mentioned as particularly potent.
44. See e.g., Mazurana and Carlson, "From Combat to Community," p. 13; David M. Crane, "Opening Statement by the Prosecutor David M. Crane against Issa Hassan Sesay, Morris Kallon and Augustin Gbao" (Freetown: Special Court for Sierra Leone, 2004); *A Journey without Map,* Brabazon (dir.).
45. Interview, civil society activist, Bo, February 2005.
46. Interviews, Kamajor ex-combatant, Freetown February 2005, civil society activist, Bo February 2005.

47. Interview, senior Kamajor commander, Kenema May 2005, see also Smith, Gambette, and Longley, "Conflict Mapping in Sierra Leone," p. 45.
48. A female ex-combatant interviewed in Mazurana and Carlson, "From Combat to Community," p. 13.
49. Interview, Kamajor ex-combatant, Freetown February 2005.
50. Interview, civil society activist, London, May 2004. A similar account was given by the Kamajor ex-combatant quoted earlier, testifying to the Special Court for Sierra Leone SC-SL, Witness TF2–021, November 2, 2004.
51. Interview, civil society activist, London May 2004; Muana, "The Kamajoi Militia," p. 89.
52. Incompetent is a pejorative term and requires some qualification in this case. What a Western soldier or observer would see as incompetent or wrong use of a weapon derives largely from the assumption that the raison d'etre of the weapon is to hit targets and that it is the responsibility of the person handling it to make sure it does by following certain rules like aiming it correctly and keeping it in good condition to avoid malfunctioning. However, if the relationship between the handler, the weapon, and the target(s) is perceived differently, "incompetence" becomes less meaningful. If the likelihood of a bullet hitting its target is seen, not as the result of the handler's skill, but as a complex interaction of supernatural power to attack countered by supernatural power to protect, the manipulation of the weapon becomes less important. Another dimension to the trial of magical power through displays to bear in mind is the relationship between the parties—those offering to have their magic tested and those offered to test it. Instances have certainly been reported where the offer is declined or the target of the test changed from a person to an animal as a gesture of respect, but this does not appear to have been the norm.
53. Interview, civil society activist, London May 2004 and Bo February 2005; see also Muana, "The Kamajoi Militia," p. 89.
54. Richards, "The Emotions at War: Atrocity as Piacular Rite in Sierra Leone."
55. Ibid.
56. Interview, senior Kamajor commander, Kenema, May 2005.
57. "Testimony Samuel Hinga Norman," (Special Court for Sierra Leone, January 27, 2006), p. 45 court transcript.
58. Ibid., p. 46 court transcript.
59. See chapter 2 for a consideration of the general role of African Traditional Religion in contemporary Africa. Geschiere, *The Modernity of Witchcraft;* Masquelier, "The Invention of Anti-Tradition."

9 The Kamajors in Battle: Magic, Tactics, and Success

1. Interview, senior Kamajor commander, Kenema, May 2005.
2. Interview, former Executive Outcomes employee who was piloting the helicopter in question, Freetown, February 2005.

3. Interview, Kamajor ex-combatant (and the Kamajor in the van), Freetown, February 2005.
4. SC-SL, Witness TF2–190, February 10, 2005, p. 45.
5. Interview, civil society activist, London, May 2004.
6. *Interview with Kenneth Kuka* (2000) (radio interview transcript, Radio Netherlands, January 21, 2000). See also Hoffman, "Like Beasts in the Bush," pp. 301–302 for similar accounts.
7. Interview, civil society activist, London May 2004, also described by a Kamajor fighter in Teun Voeten, *How De Body? One Man's Terrifying Journey through an African War* (New York: Thomas Dunne Books, 2002) p. 250. The plane remained on the landing strip in Bo for the duration of the war, but was removed to make room for a UNAMSIL base and is no longer there.
8. Interview, former Executive Outcomes employee deployed with the Kamajors, Freetown, February 2005.
9. Interview, Kamajor senior commander, Koribundu, February 2005.
10. Interviews, civil society activist, London May 2004, Kamajor ex-combatants, Freetown and Bo, May 2005. *Interview with Ex-Child Combatants* (2000) (radio interview transcript, Radio Netherlands, February 11) May 10. These are almost identical to some of the Holy Spirit Safety Precautions (nos. 3, 4, 11, 18). See Behrend, *Alice Lakwena & the Holy Spirits,* p. 47.
11. Interview, Kamajor ex-combatant, Freetown, February 2005.
12. *Interview with Ex-Child Combatants.*
13. Hoffman, "The Kamajors of Sierra Leone," p. 138. The High Priest was at one stage arrested for his initiation "business" and reprimanded by the CDF leadership.
14. For example, CDF administrators who had not been combatants were replaced with Kamajors who had combat experience to strengthen legitimacy and respect. To more directly tackle issues of tension between Kamajors and civilians and Kamajors and traditional authorities, the Office of the Director of War for the CDF engaged with civil society in civic education and reconciliation to bolster mutual respect and relieve tension.
15. In Human Rights Watch reports, by the Truth and Reconciliation Commission, the Special Court, and even on film (*A Journey without Map,* Brabazon (dir.) of Kamajors fighting in neighboring Liberia.
16. SC-SL, Witness TF2–021, November 2, 2004, pp. 70–77.
17. *A Journey without Map,* Brabazon (dir.), *The Cannibals War* (Journeyman Pictures, 1996).
18. Hoffman, "Expert Report on the Kamajors of Sierra Leone," p. 41, personal communication former Executive Outcomes employee, Freetown, May 2005.
19. Interview, former Executive Outcomes employee, Freetown, February 2005.
20. Most ex-combatants interviewed said they received no formal training but "learned on the job."
21. Iron, "Military Expert Witness Report," p. D-10.

22. Iron, "Military Expert Witness Report," p. D-11.
23. Ibid., p. D-4.
24. Interview, Kamajor senior commander Koribundu, February 2005.
25. Personal communication James Brabazon, London, April 2005.
26. Danny Hoffman, "The Civilian Target in Sierra Leone and Liberia: Political Power, Military Strategy, and Humanitarian Intervention," *African Affairs,* 103/411 (2004) p. 222.
27. Ibid.
28. Ibrahim Abdullah and Patrick K Muana, "The Revolutionary United Front of Sierra Leone: A Revolt of the Lumpenproletariat," in Christopher S. Clapham (ed.), *African Guerrillas* (Oxford: J. Curry, 1998) p. 186.
29. Although within the CDF, the various civil defense militias from around the country of course drew on regional variations of spiritual belief and practice.
30. Krijn Peters, "Footpaths to Reintegration: Armed Conflict, Youth and the Rural Crisis in Sierra Leone," PhD (Wageningen University, 2006), p. 73. Interestingly two survived, but what their role was after this trial is not revealed.
31. Ibid.
32. Interviewed by Richards, "The Emotions at War: Atrocity as Piacular Rite in Sierra Leone," p. 10 (draft). Similar accounts given in interviews to Peters, "The Storm Is Not yet Over?."
33. This was also the case in Liberia where many of the factions drew their recruits from young men disillusioned with traditional authority and sought spiritual power from outside the locality or even Liberia itself. See Ellis, *The Mask of Anarchy.*
34. See Voeten, *How De Body?,* p. 208, Ellis and Haar, *Worlds of Power,* p. 79, Interview, civil society activist, London May 2004.
35. See chapter 3 as well as Finnström, "Living with Bad Surroundings," Behrend, *Alice Lakwena & the Holy Spirits.*
36. Paul Richards, "Green Book Millenarians? The Sierra Leone War within the Perspective of an Anthropology of Religion," in Niels Kastfelt (ed.), *Religion and African Civil Wars* (London: Hurst, 2005).
37. This also featured as part of Kamajor initiation. Susan Shepler, "The Social and Cultural Context of Child Soldiering in Sierra Leone," Workshop on Techniques of Violence in Civil War (Oslo: PRIO, 2004) p. 23.
38. Many senior Kamajors have retained positions of power and influence in their communities after the war, either in local government or influential civil society organizations. Interviews, civil society activist, London May 2004 and Bo February 2005, former Executive Outcomes employee, Freetown February 2005 and civil society activist, Freetown May 2005.
39. Interview, former Executive Outcomes employee and eyewitness to the event, Freetown, February 2005.
40. Although the Kamajors fought alongside Khobe's ECOMOG troops in Sierra Leone, they received most of their arms and support from ECOMOG Liberia and other external supporters.

41. Relations between ECOMOG II (the force mandated for Sierra Leone) and the Kamajors did deteriorate although they fought together and the CDF came under the command of ECOMOG. Most notably ECOMOG were the ones to blow the whistle on the Sandline weapons shipment in 1998, preventing the supplies from reaching the CDF. In fact, both the Kamajors and the CDF more widely received the bulk of their support from the ECOMOG force in Liberia who had no reason to challenge the CDF for their turf in Sierra Leone but rather benefited from a weaker RUF that extended to affect their sponsor Charles Taylor in Liberia.
42. In the course of doing interviews for this thesis, I have had to decline offers of demonstrations for similar reasons.
43. Ellis and Haar, *Worlds of Power.*
44. Even if the power or the use of it was not approved of, its potency was acknowledged in a way similar to the approach of many civilians in northern Uganda to the alleged power of Joseph Kony. E.g., one ex-combatant, quoted by Richards and Peters, who after the war converted to Christianity, now condemned Kamajor practices as witchcraft, but did not doubt their power. Krijn Peters and Paul Richards, "'Why We Fight': Voices of Youth Combatants in Sierra Leone." *Africa: Journal of the International African Institute,* 68/2 (1998), pp. 183–210.
45. *Interview with Kenneth Kuka.*
46. A very common cause of the limited casualties in gun battles, as many young fighters had had no proper training in the use of automatic weapons and often imitated Hollywood heroes such as Rambo, spraying bullets from the hip rather than taking aim.
47. Personal communication ex-IMATT staff, London, January 2004.
48. Voltaire et al., *The Works of Voltaire: A Contemporary Version* (New York: Dingwall-Rock, 1927) Vol. IV: The Philosophical Dictionary Part 2, under entry for "Enchantment."
49. "Testimony Samuel Hinga Norman," (Special Court for Sierra Leone, January 27, 2006), p. 45 court transcript.
50. Other factors include the close community and family ties shared by the fighting sides, as well as the geographical surroundings that allowed fighters to hide effectively in the dense bush but exposed civilians in villages and other inhabited areas that were usually cleared of forest.

10 Local Practice and the Wider World of Warfare

1. See e.g., Antonio Giustozzi, *Koran, Kalashnikov and Laptop: The Neo-Taliban Insurgency in Afghanistan* (London: Hurst, 2007).
2. Hoffman, "The Meaning of Militia," pp. 639–662.
3. This is explored with reference to Sierra Leone in chapter 8.
4. See chapter 7 for a description and discussion of these as mobilizational tools.
5. In the last few years the idea of bringing anthropologists into conflict zones to assist military units in their engagement with local communities and

to "translate" the local context to the makers of strategy and tactics has been partly championed by some of the US forces in Afghanistan. However, coming to this realization halfway through a military campaign will risk seeming insincere and possibly even manipulative to the local people on the ground, rather than a genuine desire to understand. This is particularly the case with local people who have witnessed the lack of such an approach for years. As a result, the efforts risk becoming counterproductive. The approach also relies on the willingness of scholars to operate alongside and with a military organization, which is likely to complicate the objectivity seen as so central to effective academic enquiry and therefore discourage many possible candidates.

6. As discussed in chapter 5, there were other contributing factors, notably the involvement of Guinea in the tri-border area of Guinea, Sierra Leone, and Liberia, and the significant strengthening of the UN peacekeeping presence in the country. However, the British intervention took the wind out of the rebels' sails for long enough to let this combination of factors undermine their determination to keep fighting decisively.
7. US Department of Defense, *Department of Defense Dictionary of Military and Associated Terms*, April 12, 2001 (JP 3–53) p. 441.
8. See chapter 3 for a discussion of both these cases.
9. Sarah B. King, "PSYOP and Persuasion: Applying Social Psychology and Becoming an Informed Citizen," *Teaching of Psychology* 31 (January 2004).

BIBLIOGRAPHY

Abdullah, Ibrahim. "Bush Path to Destruction." *Africa Development*, XXII/3/4 (1997).

Abdullah, Ibrahim and Patrick K. Muana. "The Revolutionary United Front of Sierra Leone: A Revolt of the Lumpenproletariat." In Christopher S. Clapham (ed.), *African Guerrillas*, 172–201 (Oxford: J. Curry, 1998).

Adebajo, Adekeye. Building Peace in West Africa: Liberia, Sierra Leone, and Guinea-Bissau (International Peace Academy Occasional Paper Series; London, Boulder: Lynne Rienner, 2002).

Archibald, Steven and Paul Richards. "Converts to Human Rights? Popular Debate about War and Justice in Rural Central Sierra Leone." *Africa*, 72/3 (2002).

Ashforth, A. "An Epidemic of Witchcraft? The Implications of Aids for the Post-Apartheid State." *African Studies*, 61/1 (July 2002).

Bayart, J-F. *The State in Africa: The Politics of the Belly* (New York: Longman, 1993).

Beatty, K. J. *Human Leopards: An Account of the Trials of Human Leopards before the Special Commission Court* (London: H. Rees, 1915).

Behrend, Heike. *Alice Lakwena & the Holy Spirits: War in Northern Uganda, 1985–97* (Oxford: J. Currey, 1999).

———. "Power to Heal, Power to Kill, Spirit Possession and War in Northern Uganda (1986–1994)." In Heike Behrend (ed.), *Spirit Possession, Modernity and Power in Africa*, 20–33 (Oxford: J. Currey, 1999).

Bhebe, Ngwabi and T. O. Ranger. *Society in Zimbabwe's Liberation War* (Oxford: J. Currey, 1996).

———. *Soldiers in Zimbabwe's Liberation War* (London/Harare: J. Currey, 1995).

Bledsoe, C. H and K. M. Robey. "Arabic Literacy and Secrecy among the Mende of Sierra Leone." *Man*, New Series, 21/2 (June 1986).

Bourdieu, Pierre. *In Other Words: Essays Towards a Reflexive Sociology* (Oxford: Polity, 1990).

———. *The Logic of Practice* (Cambridge: Polity, 1990).

———. *Outline of a Theory of Practice* (Cambridge: Cambridge University Press, 1977).

Bozeman, Adda. "War and the Clash of Ideas." *Strategic Intelligence and Statecraft. Selected Essays* (Washington DC: Brassey's [US], 1992).

Briggs, Robin. *Witches and Neighbours: The Social and Cultural Context of European Witchcraft,* Second edn. (Oxford: Blackwell, 2002).

Burrows, D. "The Human Leopard Society of Sierra Leone." *Journal of the Royal African Society,* 13/50 (January 1914).

Byman, Daniel. "Remaking Alliances for the War on Terrorism." *Journal of Strategic Studies,* 29/5, 767–811, 2006.

Cline-Cole, Reginald. "Perspectives from yet Other Places, Spaces and Voices: A Commentary on Michael Watts' 'Development and Governmentality,'" *Singapore Journal of Tropical Geography* (2003).

Crane, David M. "Opening Statement by the Prosecutor David M. Crane against Issa Hassan Sesay, Morris Kallon and Augustin Gbao" (Freetown: Special Court for Sierra Leone, 2004).

Donaldson, B. A. "The Koran as Magic." *The Muslim World,* 27 (1937).

Douglas, Mary. *Risk and Blame: Essays in Cultural Theory* (London: Routledge, 1992).

"DRC: Mai Mai leader Gedeon of Manono Territory-Known 'Good Guy,' Accused Cannibal." United Nations Organization Mission in the Democratic Republic of Congo (MONUC), April 14, 2004.

Duffield, Mark. "Post-Modern Conflict: Warlords, Post-Adjustment States and Private Protection." *Civil Wars,* Spring 1/1 (1998).

"Eastern Congo Ravaged: Killing Civilians and Silencing Protests." Vol. 12 No.3(A) (Human Rights Watch, 2000).

Ellis, Stephen. "Liberia 1989–1994: A Study of Ethnic and Spiritual Violence." *African Affairs,* 94 (1995).

———. *The Mask of Anarchy: The Destruction of Liberia and the Religious Dimension of an African Civil War* (London: Hurst, 1999).

Ellis, Stephen and Gerrie Ter Haar. *Worlds of Power: Religious Thought and Political Practice in Africa* (London: Hurst, 2004).

Erickson, B. H. "Secret Societies and Social Structure." *Social Forces,* 60/1 (1981).

Evans-Pritchard, E. E. and Eva Gillies. *Witchcraft, Oracles, and Magic among the Azande* (Oxford: Clarendon Press, 1976).

Evers Rosander, Eva and David Westerlund (eds.). *African Islam and Islam in Africa: Encounters between Sufis and Islamists* (London: Hurst, 1997).

"Ex-Combatant Views of the Truth and Reconciliation Commission and the Special Court in Sierra Leone." (Freetown: The Post-conflict Reintegration Initiative for Development and Empowerment, International Center for Transitional Justice, 2002).

Farrell, Theo. "Culture and Military Power." *Review of International Studies,* 24 (1998).

Ferme, Mariane C. *The Underneath of Things: Violence, History and the Everyday in Sierra Leone* (Berkeley: University of California Press, 2001).

Ferme, Mariane C. and Danny Hoffman. "Hunter Militias and the International Human Rights Discourse in Sierra Leone and Beyond." *Africa Today* (June 2004).

Finnström, Sverker. "Living with Bad Surroundings: War and Existential Uncertainty in Acholiland, Northern Uganda." PhD (Uppsala University, 2003).

Fithen, David Caspar. "Diamonds and War in Sierra Leone: Cultural Strategies for Commercial Adaptation to Endemic Low-Intensity Conflict." PhD (UCL, 1999).

Fitz-Gerald, Ann. "Understanding Local Dynamics in Civil Wars." *Journal of Civil Wars*, 3/1 (2000).

Fukuyama, Francis. *The End of History and the Last Man* (New York: Free Press, 1992).

Garnett, J. "Strategic Studies and Its Assumptions." In K. Booth J. Baylis, J. Garnett, and P. Williams (ed.), *Contemporary Strategy: Theories and Policies*, 3–21 (London: Croon Helm, 1975).

Gberie, Lansana. *A Dirty War in West Africa: The RUF and the Destruction of Sierra Leone* (London: Hurst, 2005).

Geschiere, Peter. *The Modernity of Witchcraft: Politics and the Occult in Postcolonial Africa* (Charlottesville, VA: University Press of Virginia, 1997).

Giustozzi, Antonio. *Koran, Kalashnikov and Laptop: The Neo-Taliban Insurgency in Afghanistan* (London: Hurst, 2007).

Göle, Nilüfer. "Snapshots of Islamic Modernities." *Daedalus*, 129/1 (Winter 2000).

Gray, Colin. *Modern Strategy* (Oxford: Oxford University Press, 1999).

Hirsch, J. L. *Sierra Leone: Diamonds and the Struggle for Democracy* (Boulder, CO: Lynne Rienner, 1996).

Hoffman, Daniel. "Expert Report on the Kamajors of Sierra Leone; Prepared for the Fofana Defense Team" (Special Court for Sierra Leone, 2006).

———. "Like Beasts in the Bush: Synonyms of Childhood and Youth in Sierra Leone." *Postcolonial Studies*, 6/3 (2003).

———. "The Meaning of Militia: Understanding the Civil Defence Forces of Sierra Leone." *African Affairs*, 106/425 (2007).

Interview with Kenneth Kuka. (radio interview transcript, Radio Netherlands, January 21, 2000).

Iron, Richard. "Military Expert Witness Report on the Civil Defence Forces of Sierra Leone." (Office of the Prosecution: Special Court for Sierra Leone, 2006).

Jackson, Michael. *In Sierra Leone* (Durham, NC: Duke University Press, 2005).

Jandora, John. "War, Culture and the Interpretation of History: The Vietnam War Reconsidered." *Small Wars and Insurgencies*, 9/2 (1998).

A Journey without Map, video documentary (Camerapix, 2003), James Brabazon (dir.).

Kaldor, Mary. *New and Old Wars: Organized Violence in a Global Era* (Cambridge, MA: Polity Press, 1999).

Kaplan, Robert. "The Coming Anarchy." *The Atlantic Monthly*, February (1994).

Katzenstein, Peter J. *Cultural Norms and National Security: Police and Military in Postwar Japan* (Ithaca, NY: Cornell University Press, 1996).

Keen, David. *Conflict and Collusion in Sierra Leone* (Oxford: J. Currey, 2005).

———. *The Economic Functions of Violence in Civil Wars.* Adelphi Paper No. 320 (London: Oxford University Press for the International Institute of Strategic Studies, 1998).

King, Sarah B. "PSYOP and Persuasion: Applying Social Psychology and Becoming an Informed Citizen." *Teaching of Psychology*, 31/1 (2004).

Krige, J. D. "The Social Function of Witchcraft." In Max Marwick (ed.), *Witchcraft and Sorcery,* 263–275 (London: Penguin Books, 1982).

Kriger, Norma J. *Zimbabwe's Guerrilla War: Peasant Voices* (Cambridge: Cambridge University Press, 1992).

Lan, David. *Guns & Rain: Guerrillas & Spirit Mediums in Zimbabwe* (London: J. Currey, 1985).

Little, K. L. "The Function of 'Medicine' in Mende Society." *Man*, 48/November (November 1948).

Lord, David. *Paying the Price: The Sierra Leone Peace Process*, ed. Catherine Barnes (Accord: *An International Review of Peace Initiatives*; London: Conciliation Resources, 2000).

Masquelier, Adeline. "The Invention of Anti-Tradition: Dodo Spirits in Southern Niger." In Heike Behrend (ed.), *Spirit Possession, Modernity and Power in Africa,* 34–50 (Oxford: J. Currey, 1999).

Mazurana, Dyan and Kristopher Carlson. "From Combat to Community: Women and Girls of Sierra Leone." (Hunt Alternatives Fund, Women Waging Peace, 2004).

Melton, J. Gordon and Martin Baumann (eds.). *Religions of the World: A Comprehensive Encyclopaedia of Beliefs and Practices* (Santa Barbara, CA: ABC Clio, 2002).

Muana, Patrick. "The Kamajoi Militia: Civil War, Internal Displacement and the Politics of Counter-Insurgency." *Africa Development*, XXII/3/4 (1997).

"Nature of the Conflict." *Witness to Truth: Report of the Sierra Leone Truth and Reconciliation Commission* (Vol. 3a).

Peters, Krijn. "Footpaths to Reintegration: Armed Conflict, Youth and the Rural Crisis in Sierra Leone." PhD (Wageningen University, 2006).

———. Re-Examining Voluntarism: Youth Combatants in Sierra Leone, Monograph, 100 (Pretoria: Institute for Security Studies, 2004).

———. "The Storm Is Not yet Over? Interviews with Ex-Combatants from the War in Sierra Leone." (Unpublished typescript: Technology & Agrarian Development Group, Wageningen University & Research Centre, The Netherlands, 2002).

Peters, Krijn and Paul Richards. "'Why We Fight': Voices of Youth Combatants in Sierra Leone." *Africa: Journal of the International African Institute*, 68/2 (1998), pp.183–210.

Ploch, Lauren. "Africa Command: U.S. Strategic Interests and the Role of the U.S. Military in Africa." CRS Report for Congress, March 10, 2008.

Ranger, T. O. *Peasant Consciousness and Guerrilla War in Zimbabwe: A Comparative Study* (London: J. Currey, 1985).

Reno, William. *Warlord Politics and African States* (Boulder, CO: Lynne Rienner, 1998).

Richards, Paul. "The Emotions at War: Atrocity as Piacular Rite in Sierra Leone." In Perri 6, Susannah Radstone, Corinne Squire, and Amal Treacher (ed.), *Public Emotions,* 62–84 (London: Palgrave Macmillan, 2006).

———. *Fighting for the Rain Forest: War, Youth & Resources in Sierra Leone* (African issues; Oxford: J. Currey, 1996).

———. "Green Book Millenarians? The Sierra Leone War Within the Perspective of an Anthropology of Religion." In Niels Kastfelt (ed.), *Religion and African Civil Wars,* 119–146 (London: Hurst, 2005).

Russell, Bertrand. *Human Society in Ethics and Politics* (London: Allen and Unwin, 1954).

Shaw, Rosalind. "Cannibal Transformations: Colonialism and Commodification in the Sierra Leone Hinterland." In Henrietta L. Moore (ed.), *Magical Interpretations, Material Realities: Modernity, Witchcraft and the Occult in Postcolonial Africa,* 50–70 (London: Routledge, 2001).

Shepler, Susan. "The Social and Cultural Context of Child Soldiering in Sierra Leone." Workshop on Techniques of Violence in Civil War (Oslo: PRIO, 2004).

"Sierra Leone: Time for a New Military and Political Strategy." *Africa Report* N° 28' (Brussels: International Crisis Group, 2001).

Smith, Alison L., Catherine Gambette, and Thomas Longley. "Conflict Mapping in Sierra Leone. Violations of International Humanitarian Law from 1991 to 2002. Preliminary Edition." No Peace Without Justice [an NGO], March 2004.

Smith, M. L. R. *Fighting for Ireland?: The Military Strategy of the Irish Republican Movement* (London: Routledge, 1995).

Smith, Paul J. "Transnational Terrorism and the al Qaeda Model: Confronting New Realities." *Parameters,* Summer 2002, pp. 33–46.

Stadler, J. "Rumor, Gossip and Blame: Implications for HIV/AIDS Prevention in the South African Lowveld." *AIDS Education and Prevention,* 15/4 (August 2003).

Tétreault, Mary Ann and Robert Allen Denemark. *Gods, Guns, and Globalization: Religious Radicalism and International Political Economy* (Boulder, CO: Lynne Rienner, 2004).

Tull, Denis. "The Reconfiguration of Political Order in Africa: A Case Study of North Kivu." Hamburg African Studies (Hamburg: RFA Institut für Afrika-Kunde Hamburg, 2005).

U.S. Department of Defense, *Department of Defense Dictionary of Military and Associated Terms,* April 12, 2001.

Vines, Alex. *Renamo: From Terrorism to Democracy in Mozambique?* Second rev. edn. (London: J. Currey, 1996).

Voeten, Teun. *How De Body? One Man's Terrifying Journey through an African War* (New York: Thomas Dunne Books, 2002).

Voltaire. *The Works of Voltaire: A Contemporary Version.* Translated by William F. Fleming (New York: Dingwall-Rock, 1927).

Weigert, Stephen L. *Traditional Religion and Guerrilla Warfare in Modern Africa* (London: Macmillan, 1995).

West, Harry G. "Creative Destruction and Sorcery of Construction: Power, Hope and Suspicion in Post-War Mozambique." *Cahiers d'études Africaines* XXXVIII (3)/147 (1997).

What Future? Street Children in the Democratic Republic of Congo (New York: Human Rights Watch: 2006).

Williams, Michael C. *Culture and Security: Symbolic Power and the Politics of International Security* (London: Routledge, 2007).

Wilson, K. B. "Cults of Violence and Counter-Violence in Mozambique." *Journal of Southern African Studies,* 18/3 (1992).

"Youth, Poverty and Blood: The Lethal Legacy of West Africa's Regional Warriors" (New York: Human Rights Watch, 2005).

INDEX